European Invertebrate Survey

Atlas of the Bumblebees of the British Isles

Bombus and *Psithyrus*
(Hymenoptera: Apidae)

Compiled by
International Bee Research
Association, London and the
Biological Records Centre,
Institute of Terrestrial Ecology,
Natural Environment Research
Council, Monks Wood Experimental
Station, Huntingdon.

1980

Printed in England by
Staples Printers St Albans Limited at The Priory Press
© Copyright 1980
Published in 1980 by
Institute of Terrestrial Ecology
68 Hills Road
Cambridge
CB2 1LA

Also obtainable from
IBRA, Hill House, Gerrards Cross
Bucks SL9 0NR

ISBN 0 964282 32 5

The photograph on the cover shows a worker of *Bombus lucorum*, taken by F. G. Vernon.

The Institute of Terrestrial Ecology (ITE) was established in 1973, from the former Nature Conservancy's research stations and staff, joined later by the Institute of Tree Biology and the Culture Centre of Algae and Protozoa. ITE contributes to and draws upon the collective knowledge of the fourteen sister institutes which make up the *Natural Environment Research Council*, spanning all the environmental sciences.

The Institute studies the factors determining the structure, composition and processes of land and freshwater systems, and of individual plant and animal species. It is developing a sounder scientific basis for predicting and modelling environmental trends arising from natural or man-made change. The results of this research are available to those responsible for the protection, management and wise use of our natural resources.

Nearly half of ITE's work is research commissioned by customers, such as the Nature Conservancy Council who require information for wildlife conservation, the Forestry Commission and the Department of the Environment. The remainder is fundamental research supported by NERC.

ITE's expertise is widely used by international organisations in overseas projects and programmes of research.

The International Bee Research Association (IBRA) was founded in 1949 for advancing research on bees and apiculture. One of its most important activities is to act as a clearing house for scientific and technical information about all species of bees, their products, their husbandry, their behaviour, and their place in nature. This work is funded by subscriptions, sales and donations, and by grants from the Development Commission, the Commonwealth Agricultural Bureaux, and other sources.

IBRA has members and subscribers in 103 countries, who recognize it as the source of scientific information on topics related to bees, and as the centre through which connection can be made with any aspect of bee research and the people interested in it. At its headquarters, Hill House, Chalfont St. Peter, Gerrards Cross, Bucks., the Association maintains one of the most comprehensive libraries of its kind in the world.

IBRA publishes three quarterly journals, *Bee World, Apicultural Abstracts* and *Journal of Apicultural Research*. It also publishes books and pamphlets, including authoritative textbooks and reprints of scientific papers, and audiovisual educational material. Details of publications and membership are available free of charge.

Introduction

BUMBLEBEE DISTRIBUTION MAPS SCHEME

For several centuries Britain has been well served by her naturalists, who have pioneered many studies of the plants and animals in their surroundings. One type of study which has been extended greatly in recent decades is the production of distribution maps of plants and animals, throughout the biogeographical region known as the British Isles.

In 1969 the European Invertebrate Survey was set up to encourage distribution mapping of Invertebrates throughout Europe. The mapping of bumblebees was carried out as part of the work of the Survey.

Bumblebees command an interest which goes far beyond that of the professional biologist and specialist. This interest is fostered by the fact that bumblebees as a group are easily recognisable, although some species are difficult to identify. There are two genera—*Bombus* and *Psithyrus* (the cuckoo bees). The survey showed there were 18 species of *Bombus* in Britain and 14 in Ireland. It also recorded 6 species of *Psithyrus* in Britain and 5 in Ireland.

The enormous task of mapping the distribution was made possible only with the help and co-operation of amateur naturalists, who acted as collectors, organised by IBRA as well as professional biologists.

The Bumblebee Distribution Maps Scheme (BDMS) started in early 1970 in European Conservation Year, which undoubtedly encouraged a wide and interested response. Entomologists who were competent to identify bumblebees were registered as Recorders; other naturalists as Collectors, who collected and sent in specimens for identification. These specimens were identified by Dr D V Alford in Cambridge, working voluntarily and in his spare time. The Recorders sent specimens with their report cards only when they were unsure of the identification. Quite a few people who started as Recorders found that they had to change later to Collector status.

Throughout the Scheme, Recorders with access to museum collections and libraries completed record cards for the collections and for references from past literature.

The Biological Records Centre of the Institute of Terrestrial Ecology (Natural Environment Research Council) processed the data from the completed record cards. The maps were printed using a specially modified electric typewriter. Two symbols are used, to differentiate the date of records: a black spot for records from 1960 onwards, and an open circle for records before 1960.

The first maps were published after only two seasons' work: one showed the 10-km grid squares from which records had been received (the earliest version of Map 1), and two showed squares from which *Bombus pascuorum* (formerly *agrorum*) and *Bombus lucorum* had been recorded. By early summer 1973 (three years after the scheme started), provisional maps had been prepared for all species, and Dr Alford had completed a series of papers on the identification of bumblebees. The map showing squares from which records had been received was published in reverse, as a "target area map" for 1974. Recording continued until 1975, ie for 6 full seasons, and a few records were received in 1976.

The number of records received greatly exceeded expectations and are quite widely spread over various parts of Britain and Ireland. It is hardly surprising that there is no information from remote areas, eg records from the boglands of Ireland are usually few. Many thousands of records were received by 1976 when the scheme was closed. These related to 2,700 10-km grid squares and resulted in some 13,500 map entries.

The maps show that today the only generally common species are *Bombus hortorum, B. lapidarius, B. lucorum, B. pascuorum, B. pratorum* and *B. terrestris*. The parasitic *Psithyrus* species, sometimes known as cuckoo bees, are never as numerous as their respective host species of *Bombus*. Several *Bombus* species (notably *humilis, ruderatus* and *sylvarum*) are now much scarcer than formerly, and more limited in their distribution (Sladen, 1912). No records have been obtained of *B. cullumanus*, and this species, like *B. pomorum*, is probably now extinct in Britain as well as in Ireland. However, on the positive side, we can report *B. lapponicus* in Ireland; Dr M C D Speight obtained the first Irish examples in 1974, as a direct result of BDMS.

Details of the biology, habits and identification of all the species represented in the maps were published in Alford (1975) and his nomenclature is followed in the maps.

The record cards give further information than that published in the maps. This includes date and locality and sometimes altitude, habitat (often with the plant species on which the bee was recorded), and whether the specimen was a queen, male or worker. The original record cards are stored at the International Bee Research Association, Gerrards Cross, and the 'Master' cards, ie the summary cards for each 10-km square, at the Biological Records Centre, Monks Wood Experimental Station.

Acknowledgements

The compilers of this atlas wish to thank all concerned for their participation in the Bumblebee Distribution Maps Scheme. Dr Eva Crane, Director of IBRA, was awarded a grant (GR3/1589) from the Natural Environment Research Council from 1971 to 1975 in aid of the BDMS investigation. Audrey Batchelor (IBRA staff) sent out instructions and information to Recorders and Collectors and handled the correspondence and address records.

Dr David Alford received all records and specimens from participants, made all the identifications required, and passed the resultant data on to the Biological Records Centre. He also checked Recorders' identifications when so requested. He handled some tens of thousands of specimens and made the appropriate entries on the record cards. All of this was done voluntarily, in his own time. Mr John Heath was in charge of data processing and map production at the Biological Records Centre.

There were 980 observers in the Scheme, 350 Recorders and 630 Collectors. Although the majority are not individually named, each one played an essential part in the production of the maps published here. Among the Recorders who made notable contributions were the late Lt. Col. W E Almond, M A Archer, H J Berman, W K Booker, I C Christie, P R Cobb, Dr G H L Dicker, H T Eales, A G Eames, J Felton, J A Lade, E G May, K A Moseley, E G Philp, G W Spooner.

The following Collectors also deserve special mention: R W Burton, R M Duggan, Professor A J Haddow, R Hewson, B G Hinton, J G Keylock, R T McAndrew, J C C Tyssen and N Young and his colleagues at the Welsh Plant Breeding Station.

Many Field Centres, Nature Reserves and Education Authorities integrated bumblebee recording into their general programmes. BDMS gained a number of additional observers, especially in less well recorded areas in Scotland, when the five-year period of field recording for the Bird Atlas of the British Trust for Ornithology ended in 1972. Among our greatest helpers were the late A. Macdonald and B P Pickess of the Royal Society for the Protection of Birds.

The Irish Biological Records Centre played an active part in the Scheme under Eanna Ni Lamhna, and the degree of completeness of the Irish records is largely due to Miss Ni Lamhna's enterprise in getting co-operation from both adults and schoolchildren. Others who made notable contributions were Dr M Speight in Dublin and R Edwards and L Forsyth in Belfast.

Among the most active contributors from museums and libraries were G Else, British Museum (Natural History); C O'Toole, University Museum, Oxford; Mrs E Smith, Royal Scottish Museum, Edinburgh; R A Holmes, Merseyside County Museum; W A Ely, Rotherham Museum; G Halfpenny, City Museum, Stoke-on-Trent and R K Merrifield, Truro Museum. The museums at Bolton, Doncaster, Dundee, Letchworth, Maidstone, Newcastle-on-Tyne and Warwick also sent material.

The International Bee Research Association much appreciates the contributions of John Heath of the Biological Records Centre to the present atlas and his co-ordination of the important series of atlases prepared under his editorship.

The idea of recording and mapping bumblebees, and of the part these bees can play in pollination, captured the imagination of a number of reporters working for the press, radio and television. We are greatly indebted to the publicity they gave to the Scheme, which brought it to the notice of many more people than could be approached directly; this, in turn, brought new recruits.

It is encouraging to us that there is so much interest in bumblebees among the general public. The Scheme obviously served a further useful function in enabling many more people to learn about bumblebees.

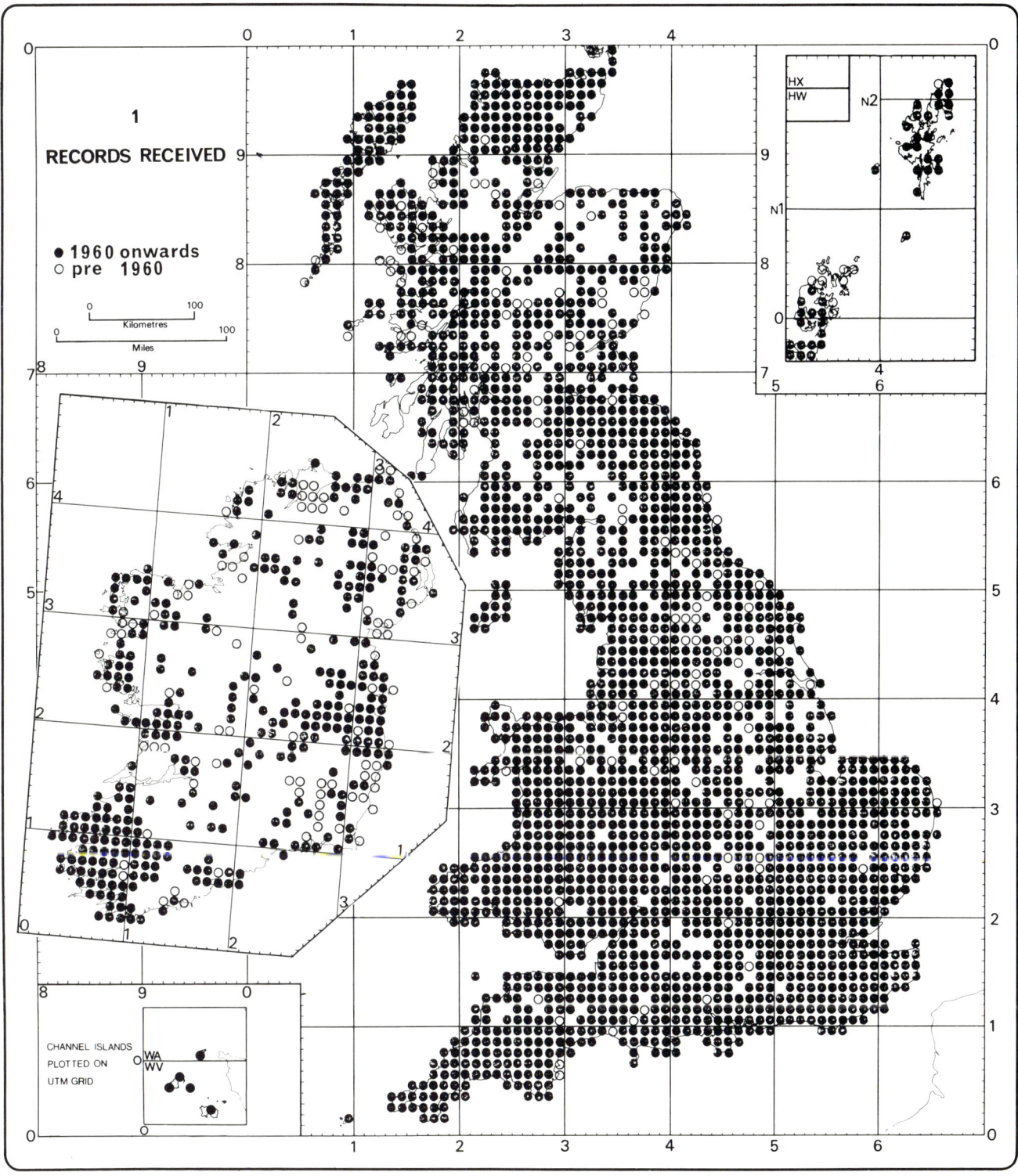

Map 1 Records received.

On all maps a 10-km square marked with an open circle represents a record before 1960 but not since.

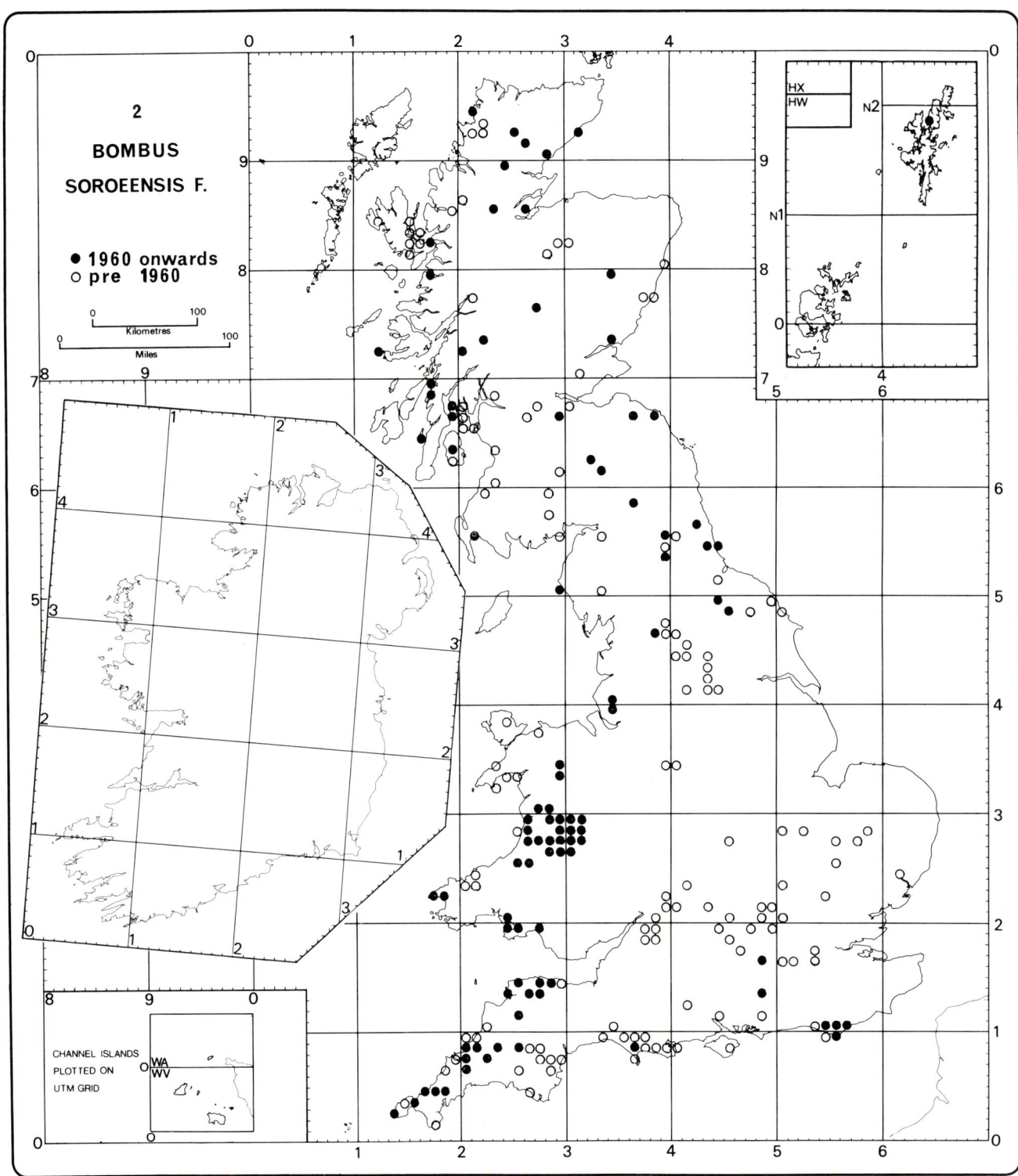

Map 2   Bombus soroeensis *(Fabr.) is always regarded as local and uncommon.*

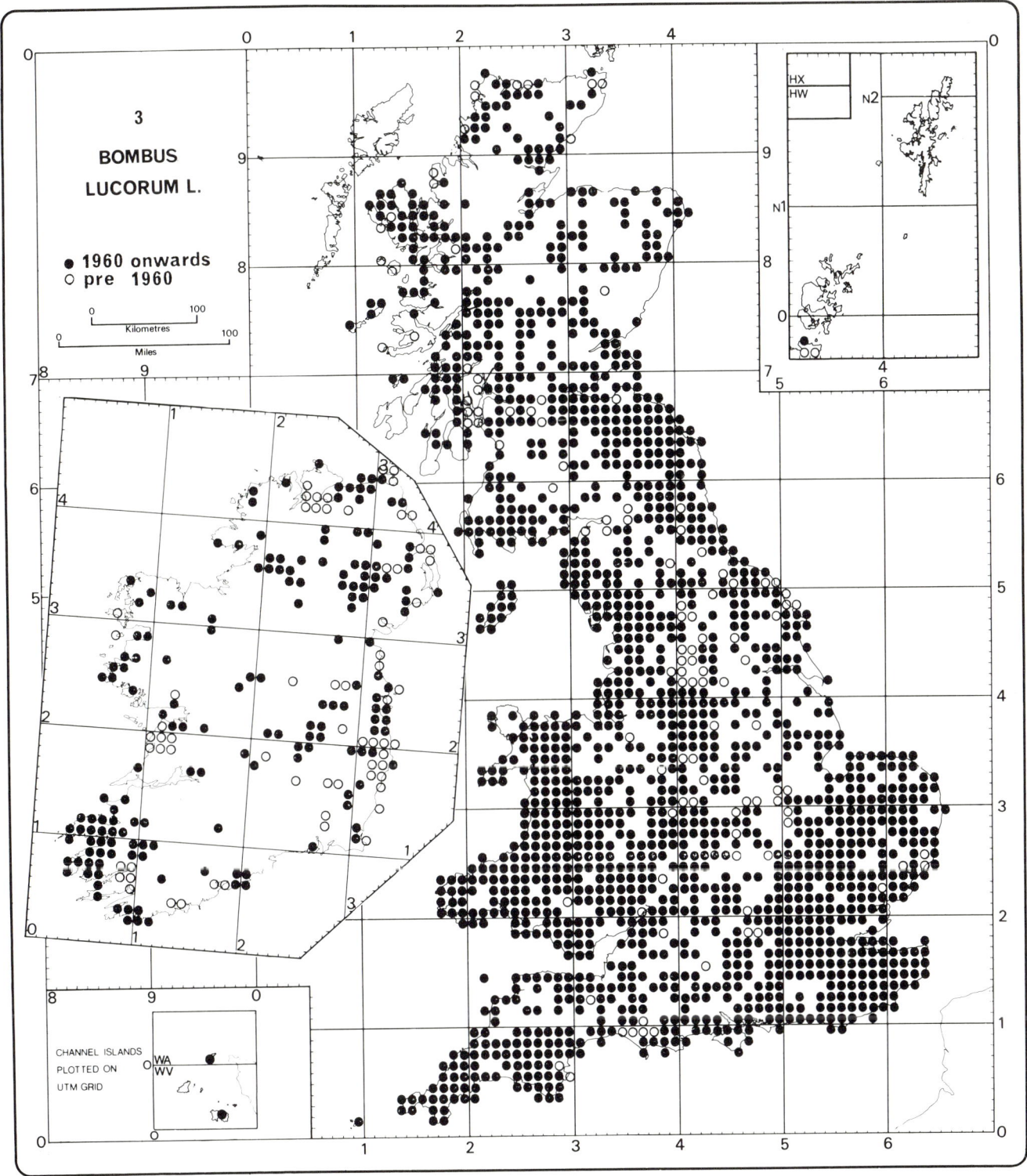

Map 3 Bombus lucorum *(L.) is one of the most abundant and widely distributed species.*

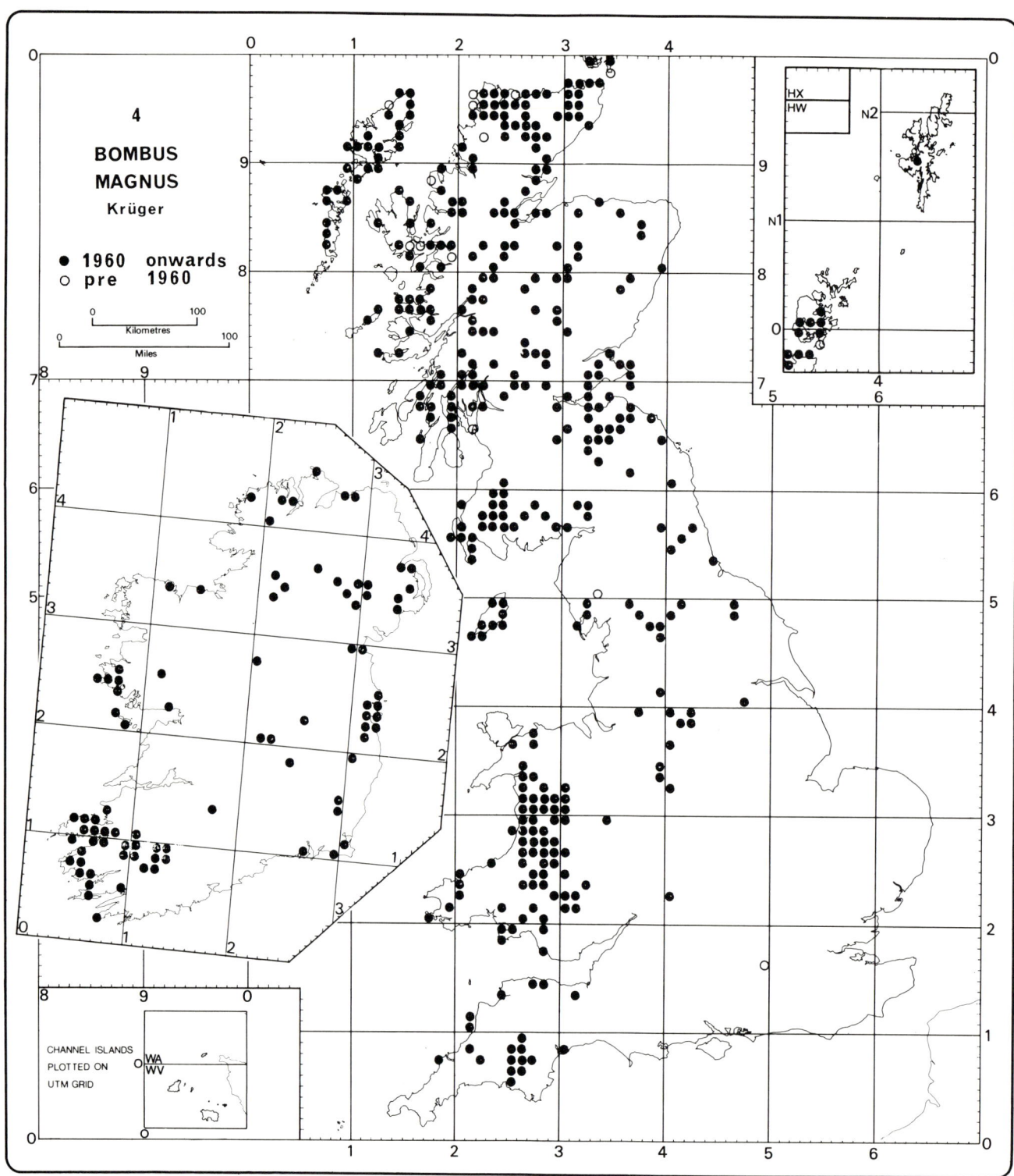

Map 4 Bombus magnus Krüger *is a close relative of* B. lucorum, *but occurs in more exposed habitats.*

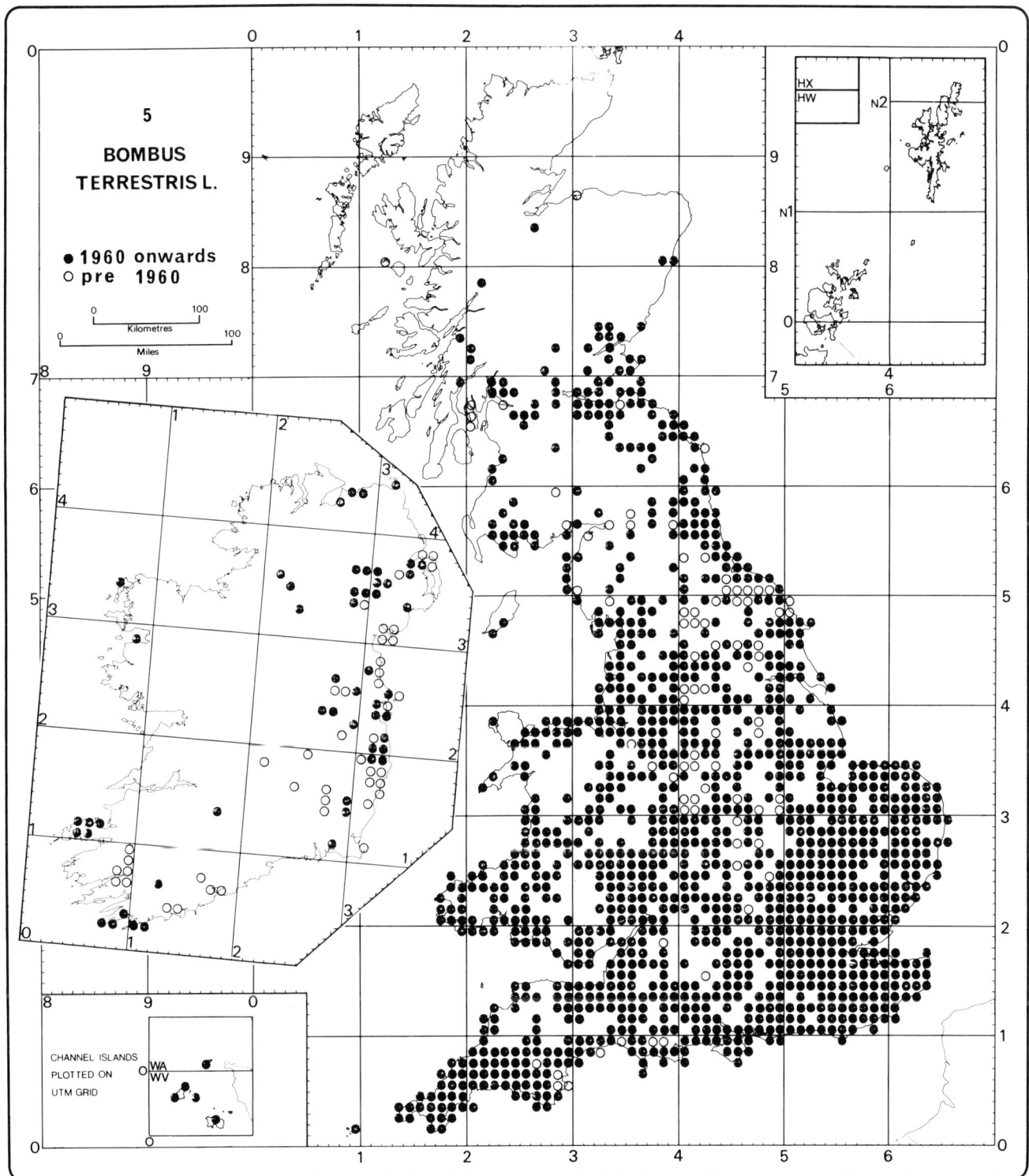

Map 5   Bombus terrestris *(L.) is one of the most common species in southern Britain.*

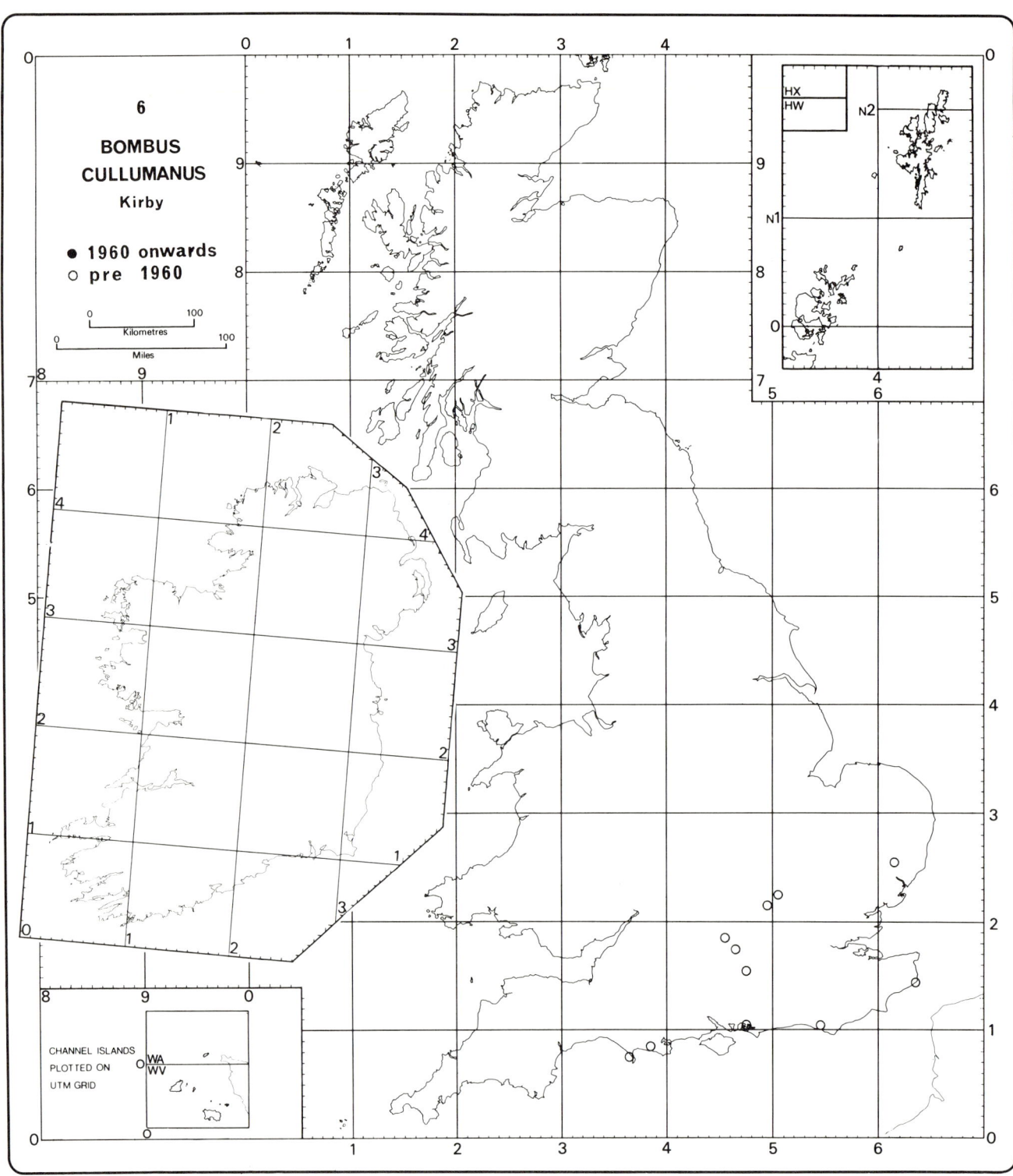

Map 6   Bombus cullumanus *(Kirby) was formerly recorded in a few chalkland sites in southern England, but is now probably extinct in Britain. It has not been recorded from Ireland.*

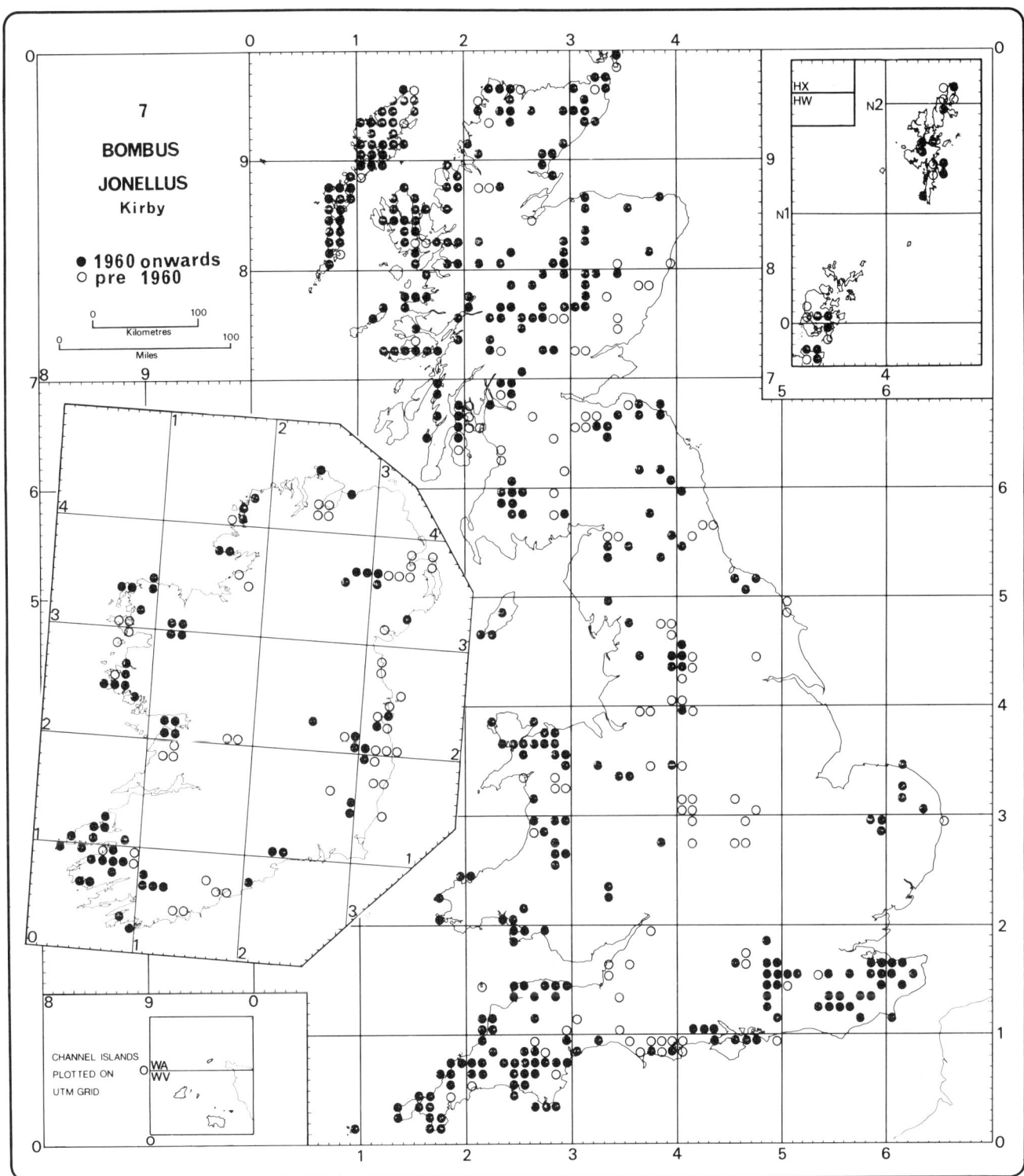

Map 7   Bombus jonellus *(Kirby)* is mainly found on heathlands and moorlands.

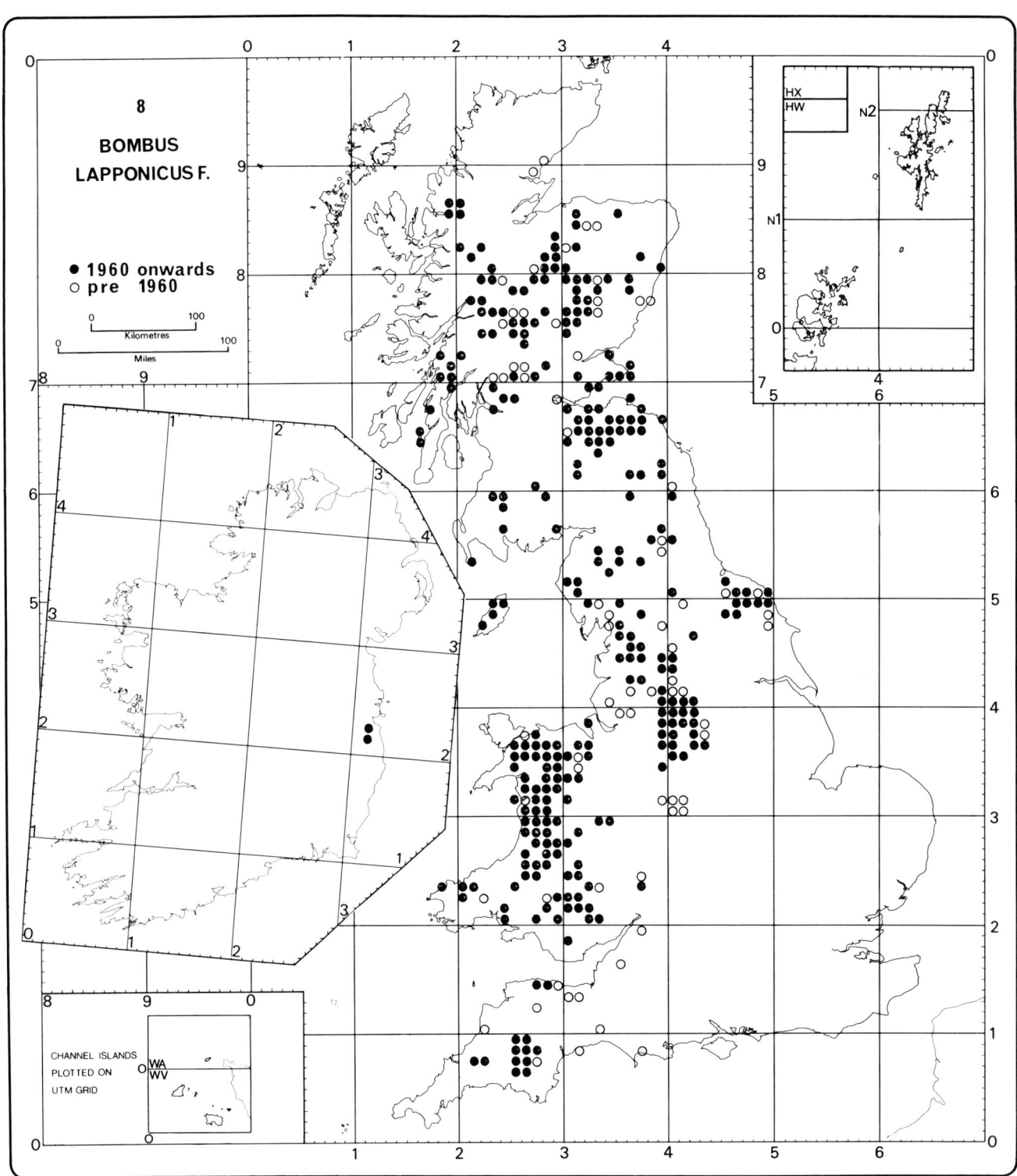

Map 8 Bombus lapponicus *(Fabr.) is associated with* Vaccinium, *it was first discovered in Ireland by a BDMS Recorder.*

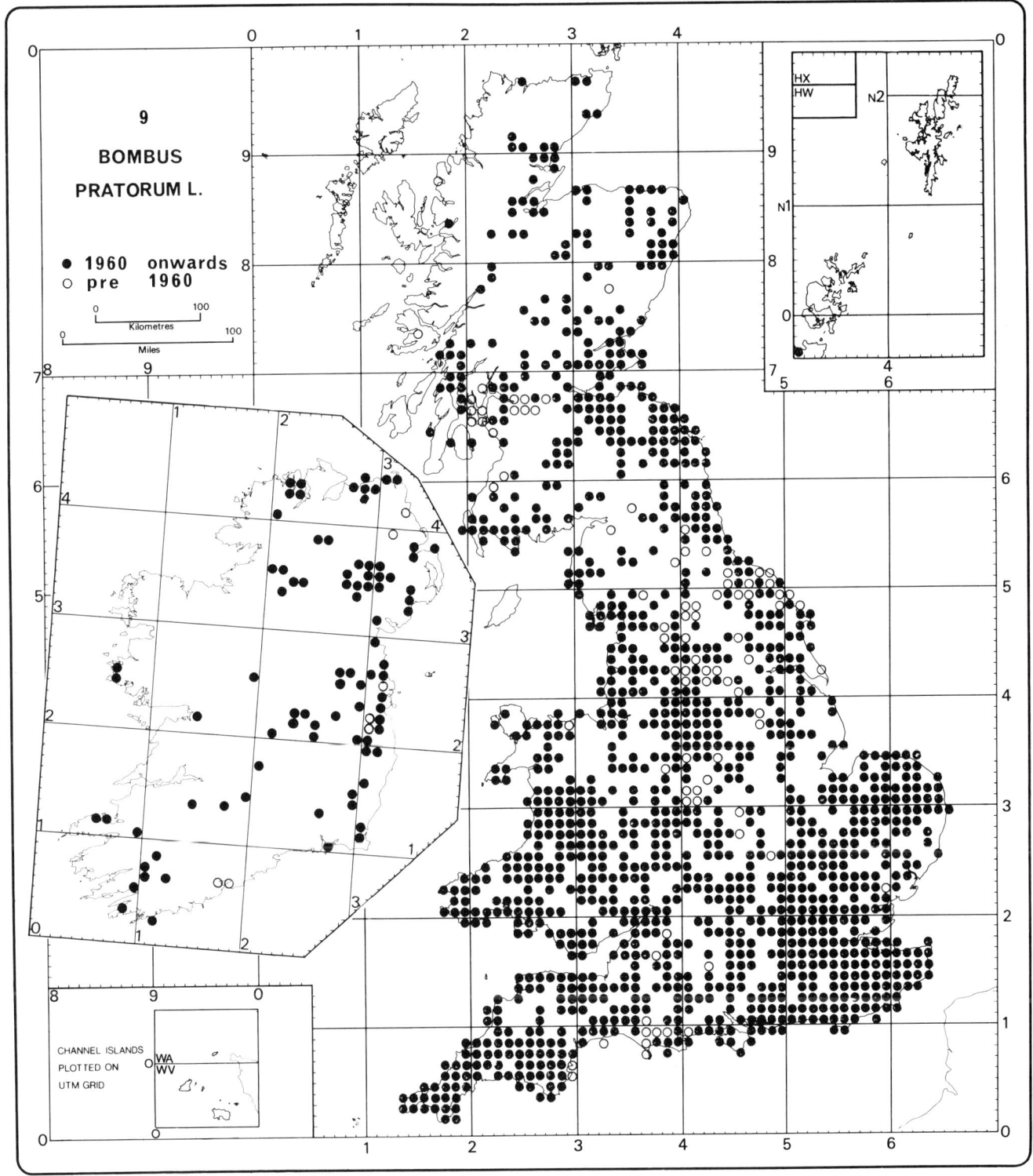

Map 9   Bombus pratorum *(L.) is widely distributed and often abundant; it was once rare in Ireland but is now relatively common there.*

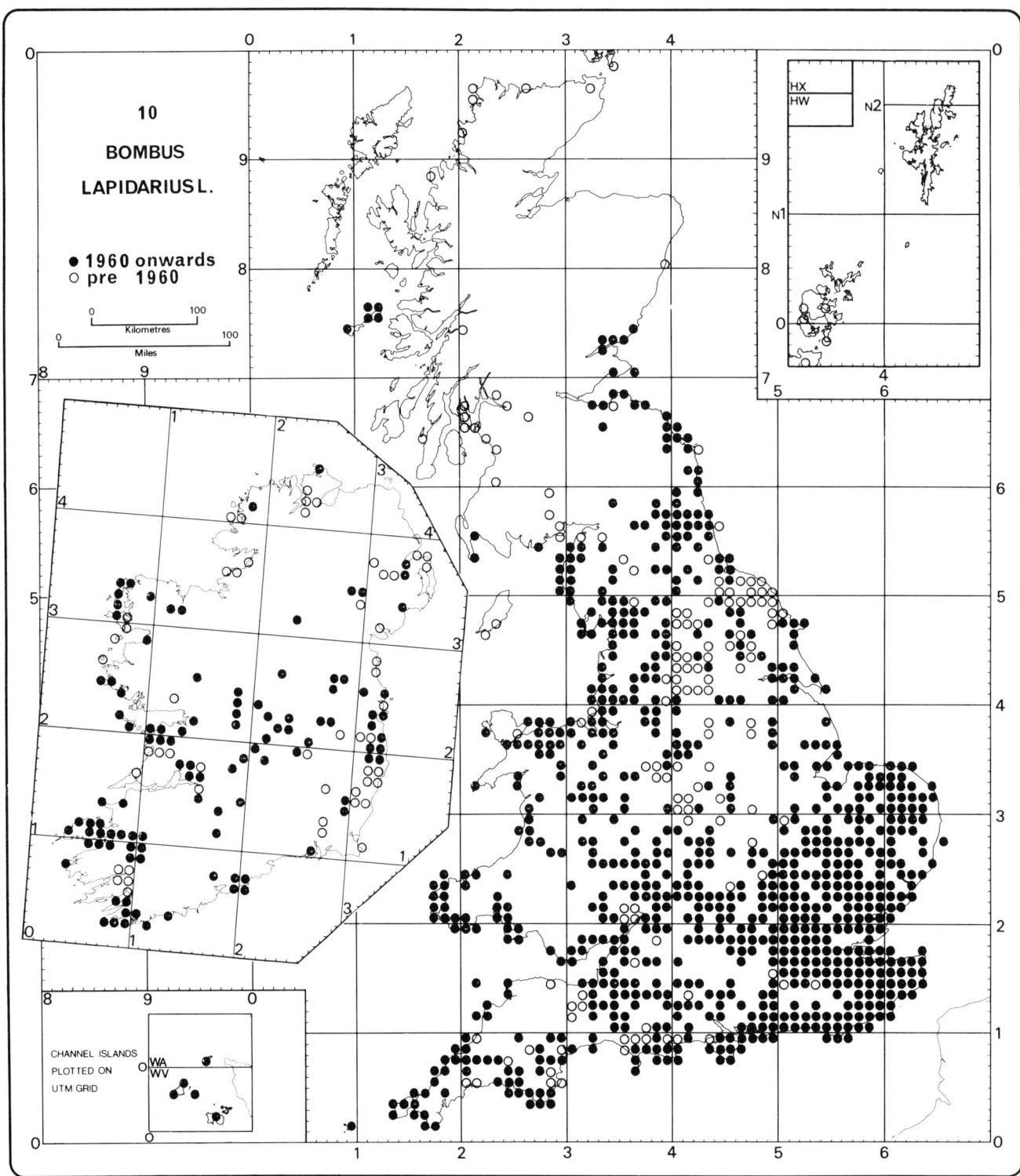

Map 10 Bombus lapidarius *(L.) is widely distributed but varies in abundance; it is often very numerous in chalkland and coastal areas.*

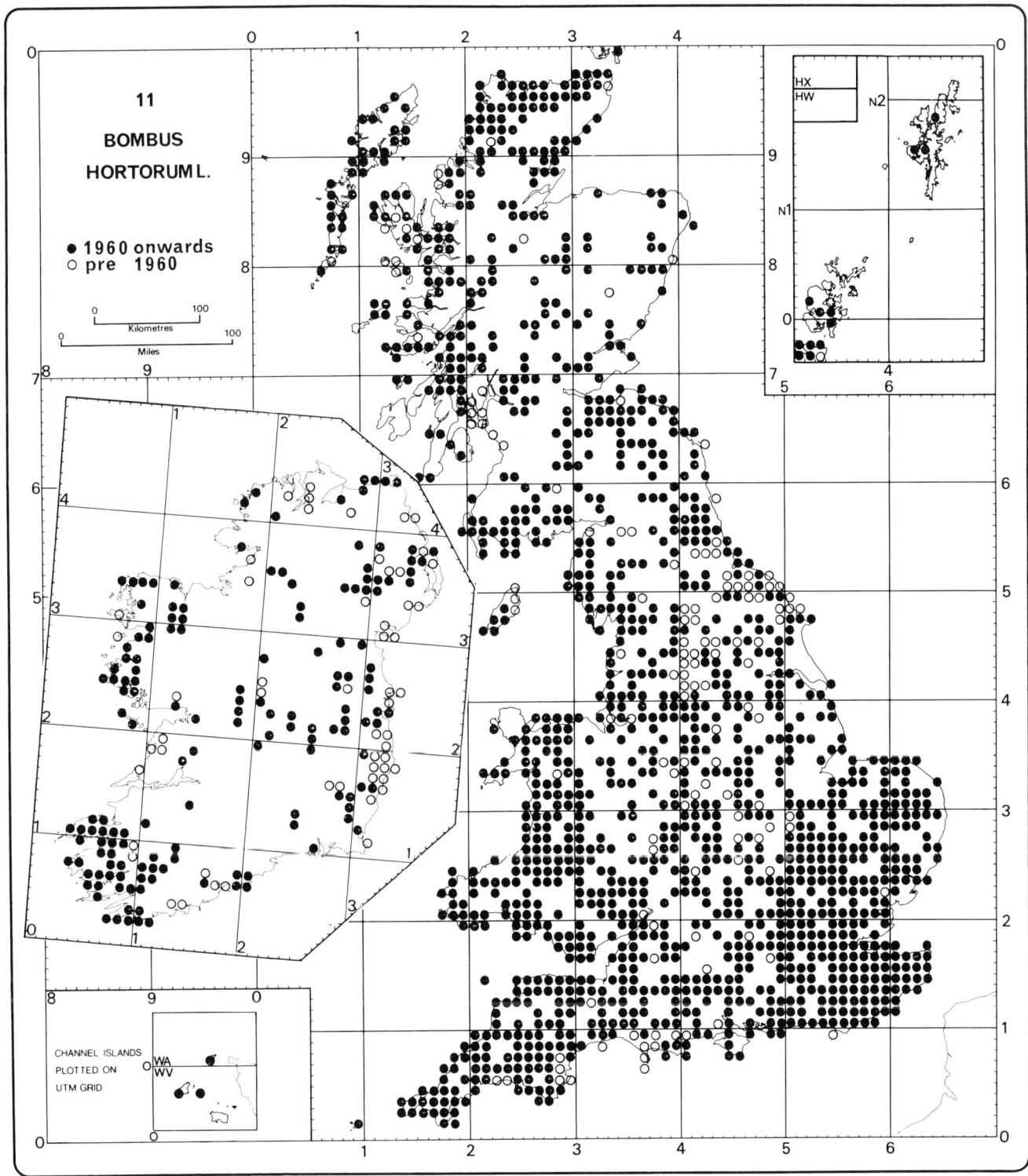

Map 11 Bombus hortorum (L.) is one of the most widely distributed and common species.

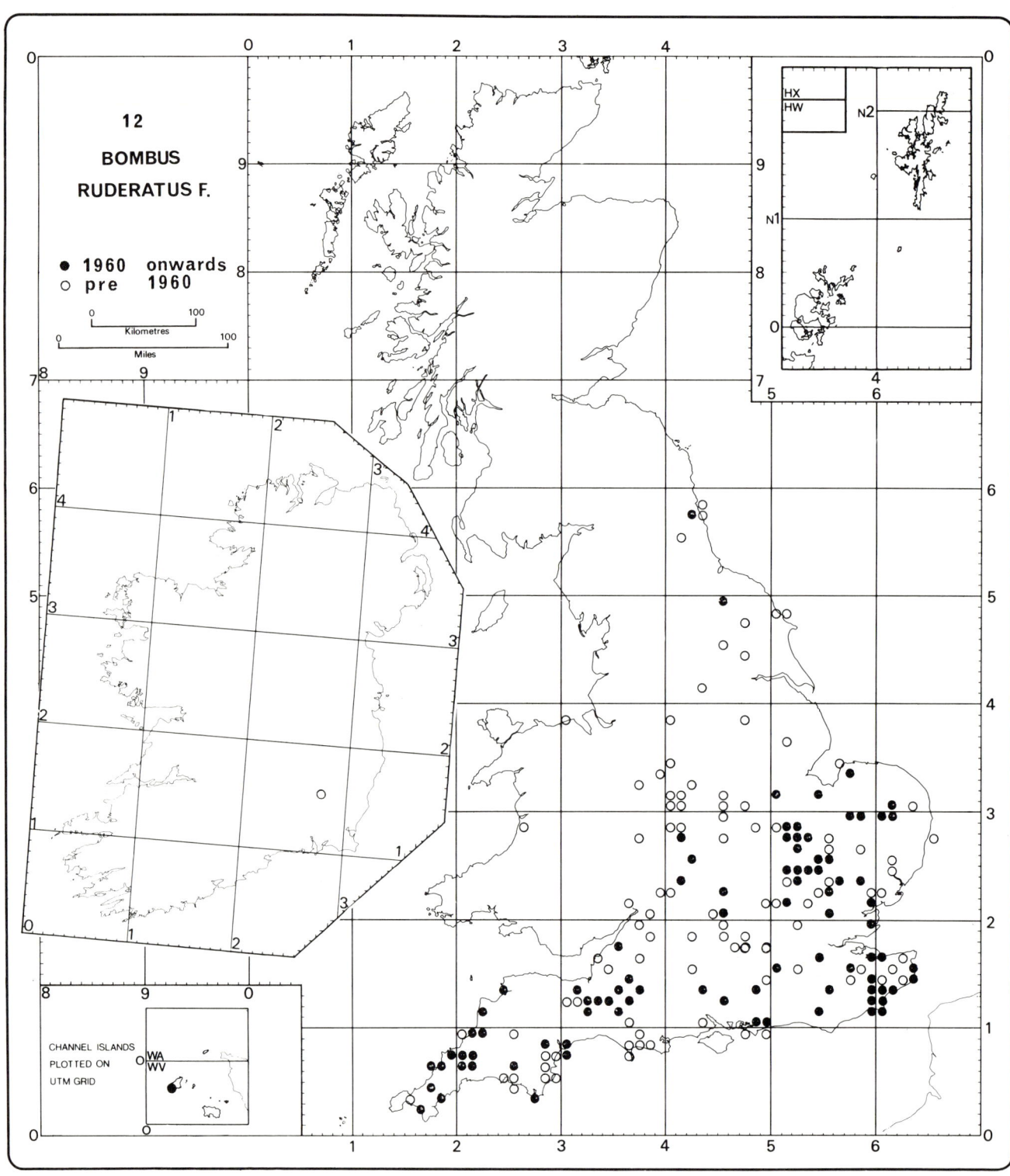

Map 12 Bombus ruderatus *(Fabr.), is a close relative of* B. hortorum, *is far less common and only locally distributed.*

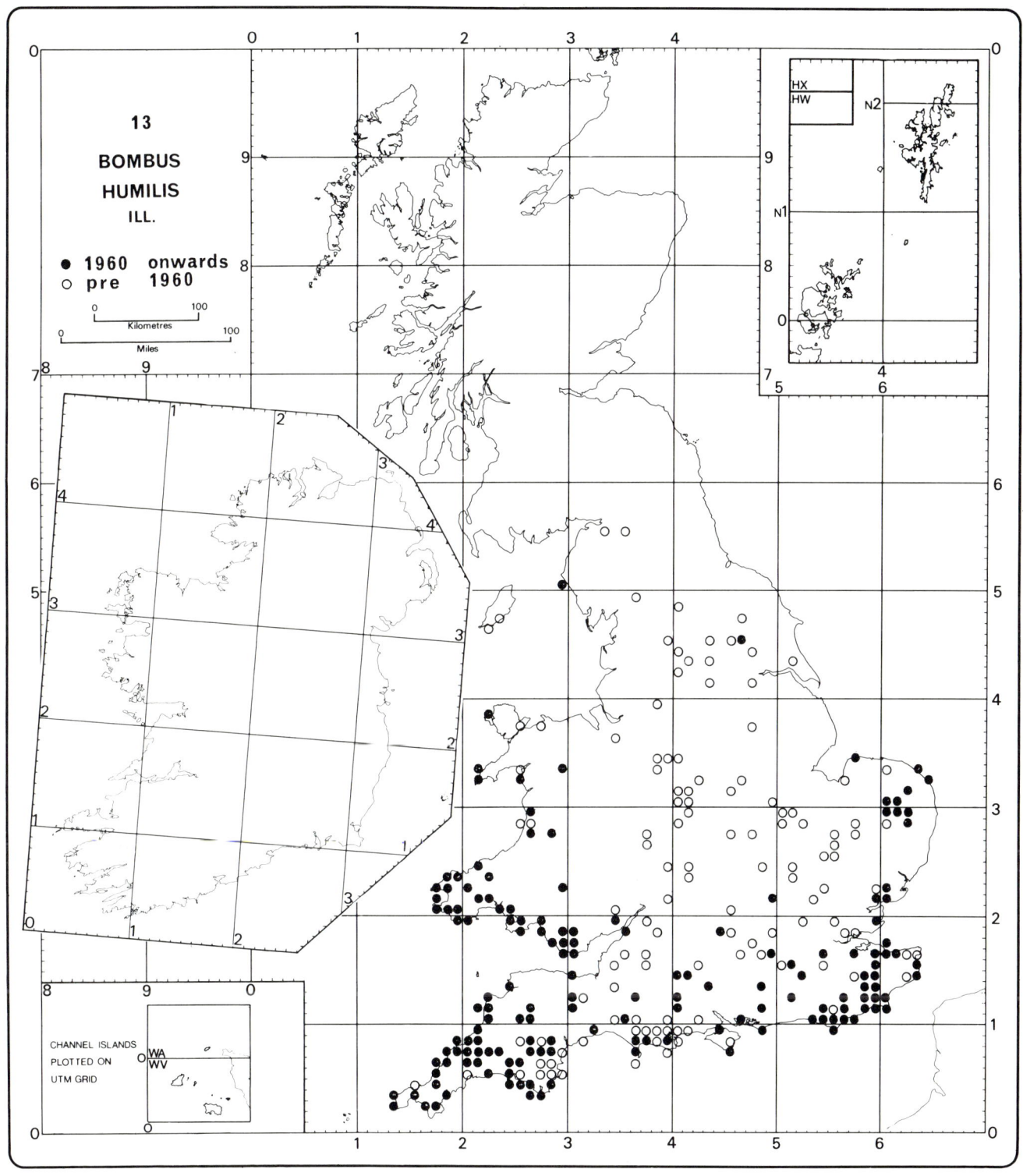

Map 13 Bombus humilis *Illiger is a local species, whose distribution shows a marked decline.*

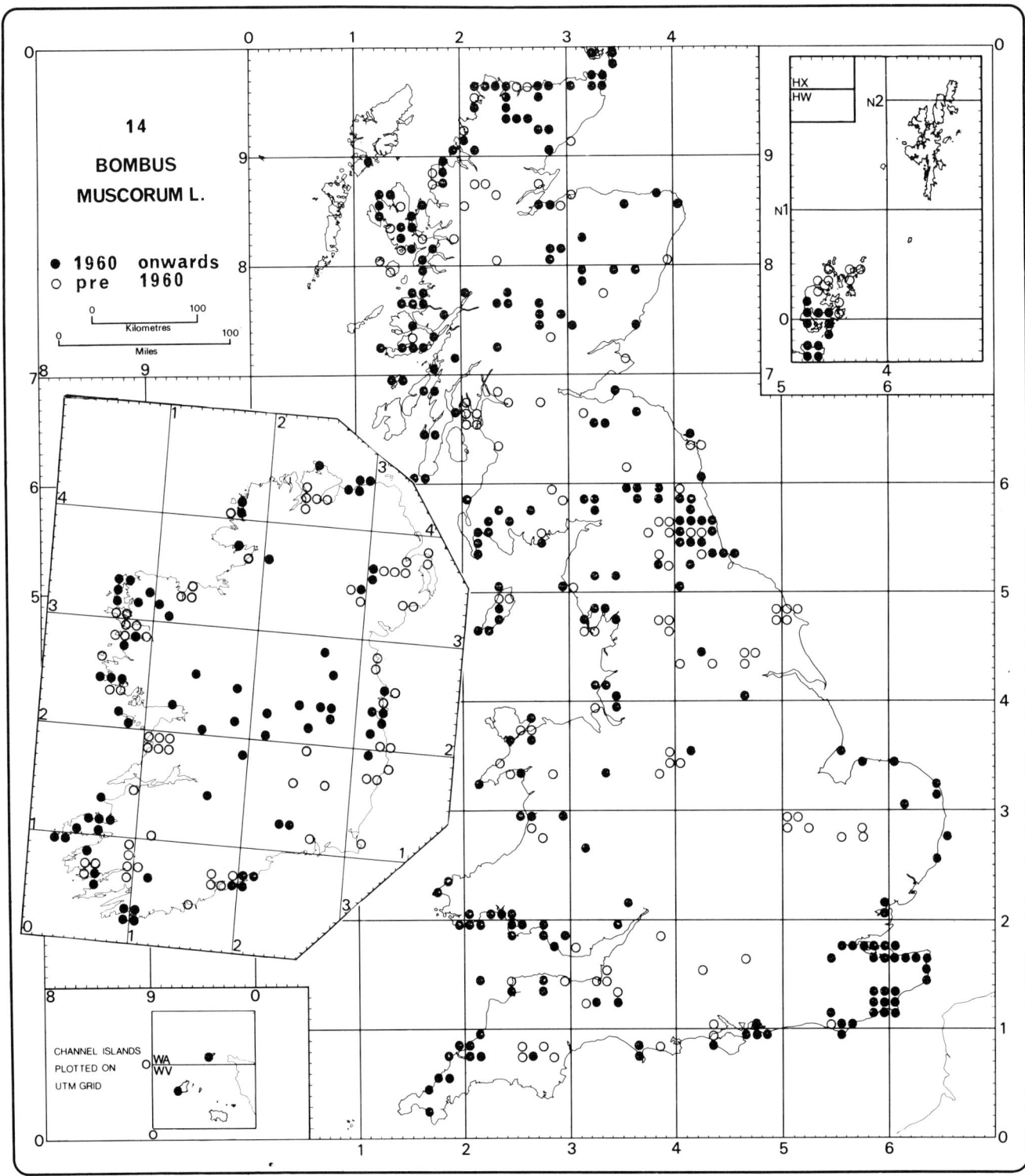

Map 14   Bombus muscorum (L.) occurs as several sub-species, in coastal, marshy or moorland areas; it is most common in the north and west of Britain.

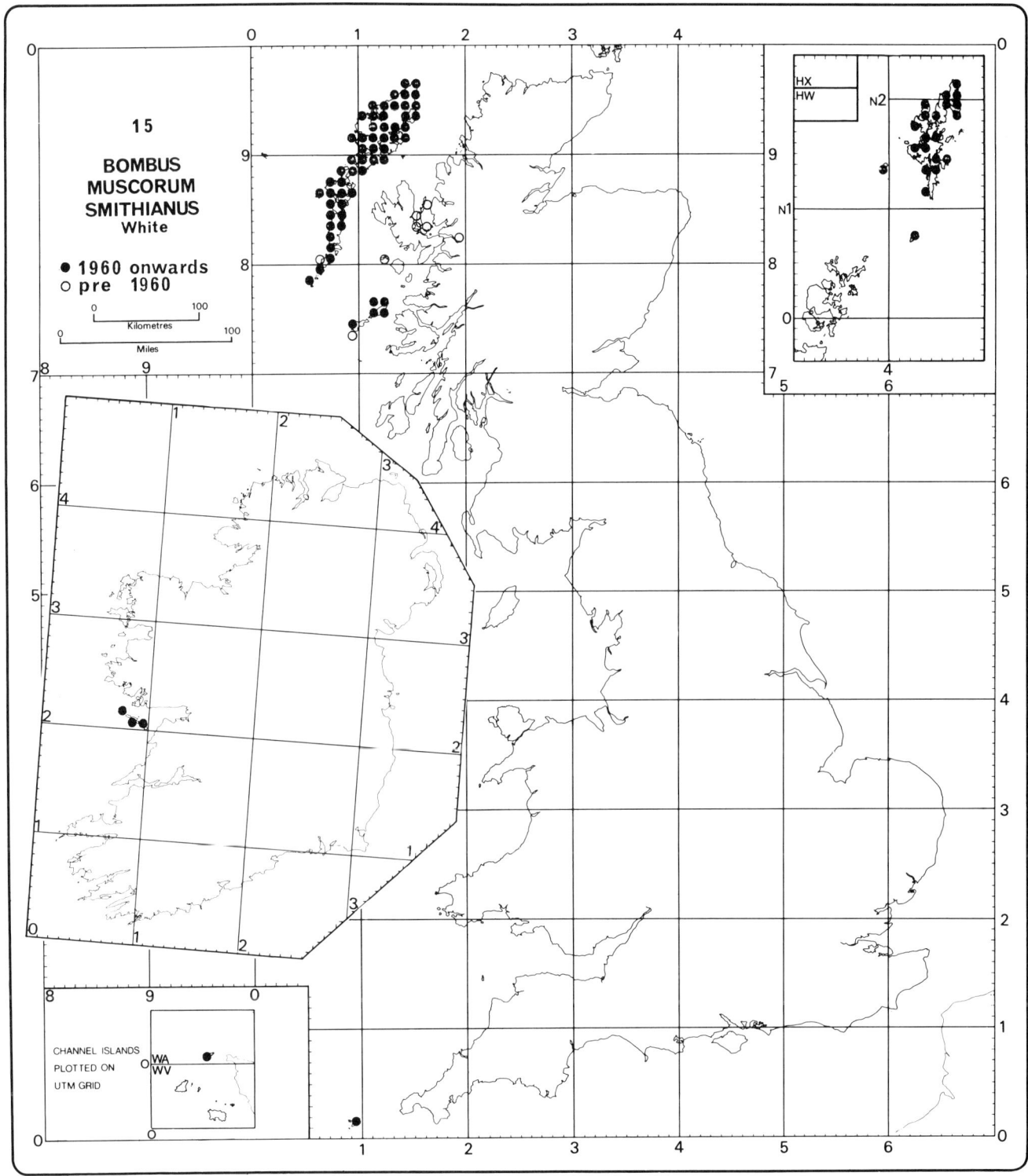

Map 15 Bombus muscorum smithianus *White includes a group of sub-species that are especially distinctive, occurring as different forms on different islands, but not on the mainland.*

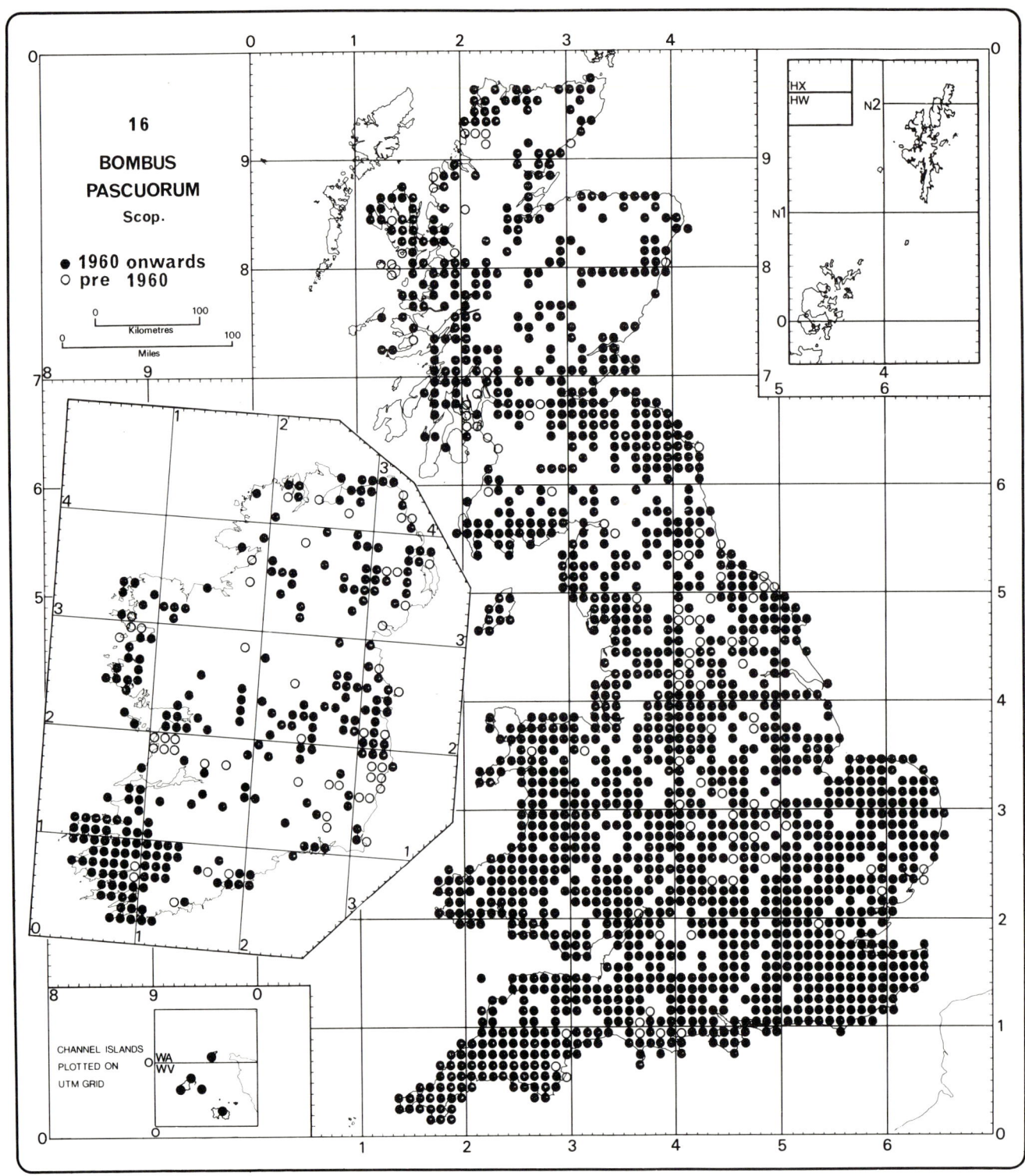

Map 16 Bombus pascuorum *(Scopoli) is one of the most abundant and widely distributed species.*

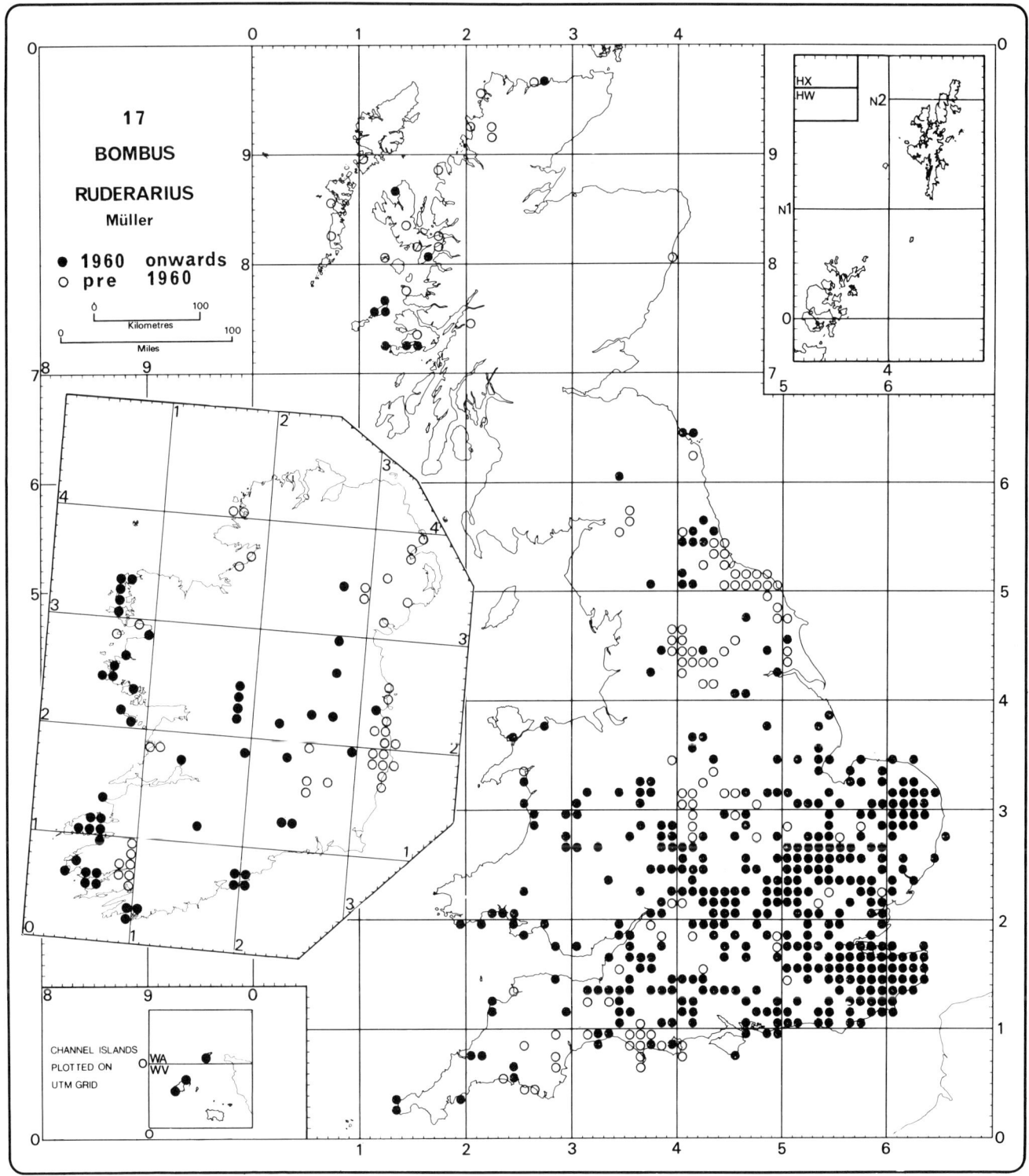

Map 17 Bombus ruderarius *(Müller)* has a rather patchy distribution: it is abundant in several areas but local or rare in others.

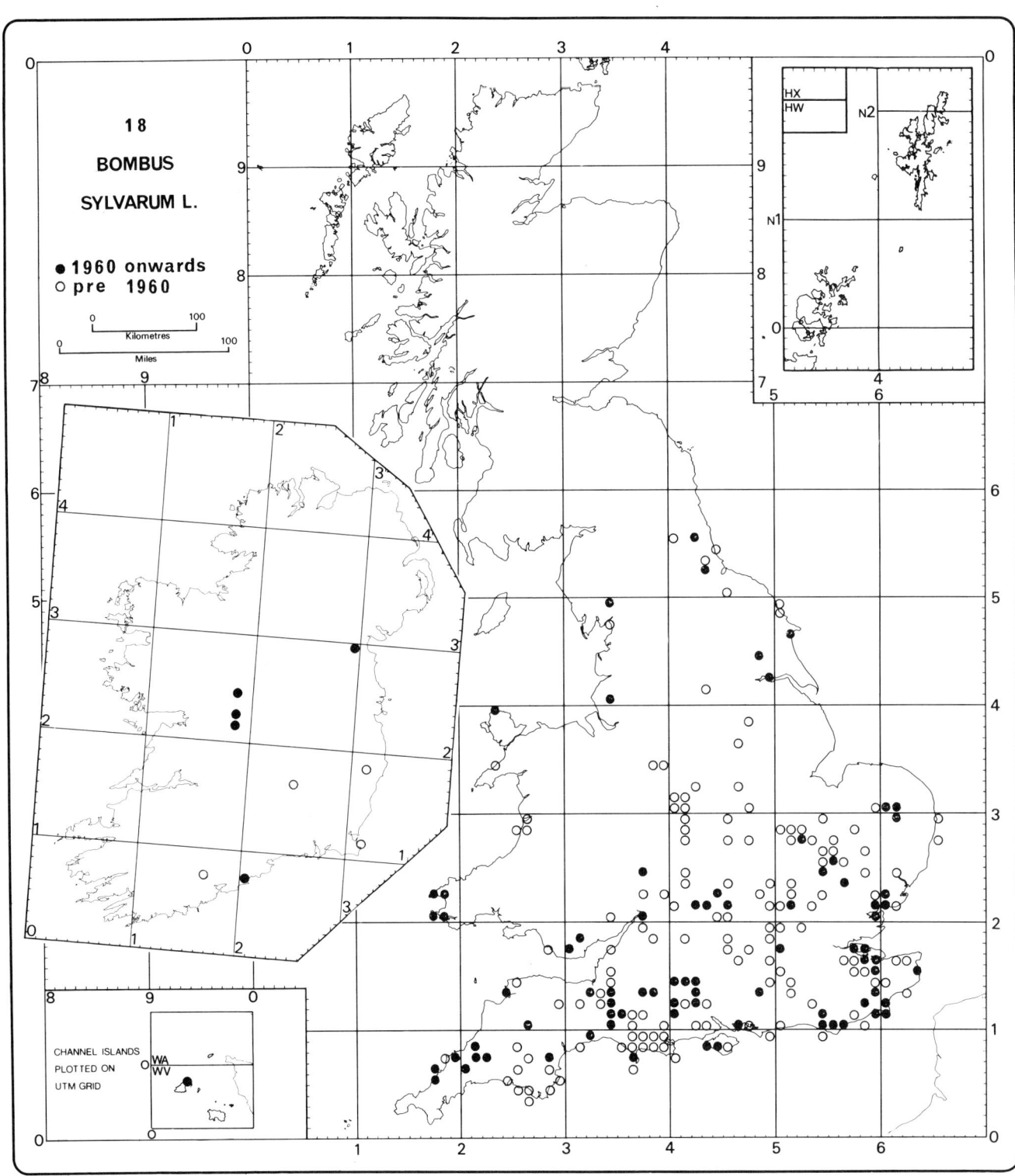

Map 18 Bombus sylvarum *(L.) is an uncommon and local species in Britain and very rare in Ireland.*

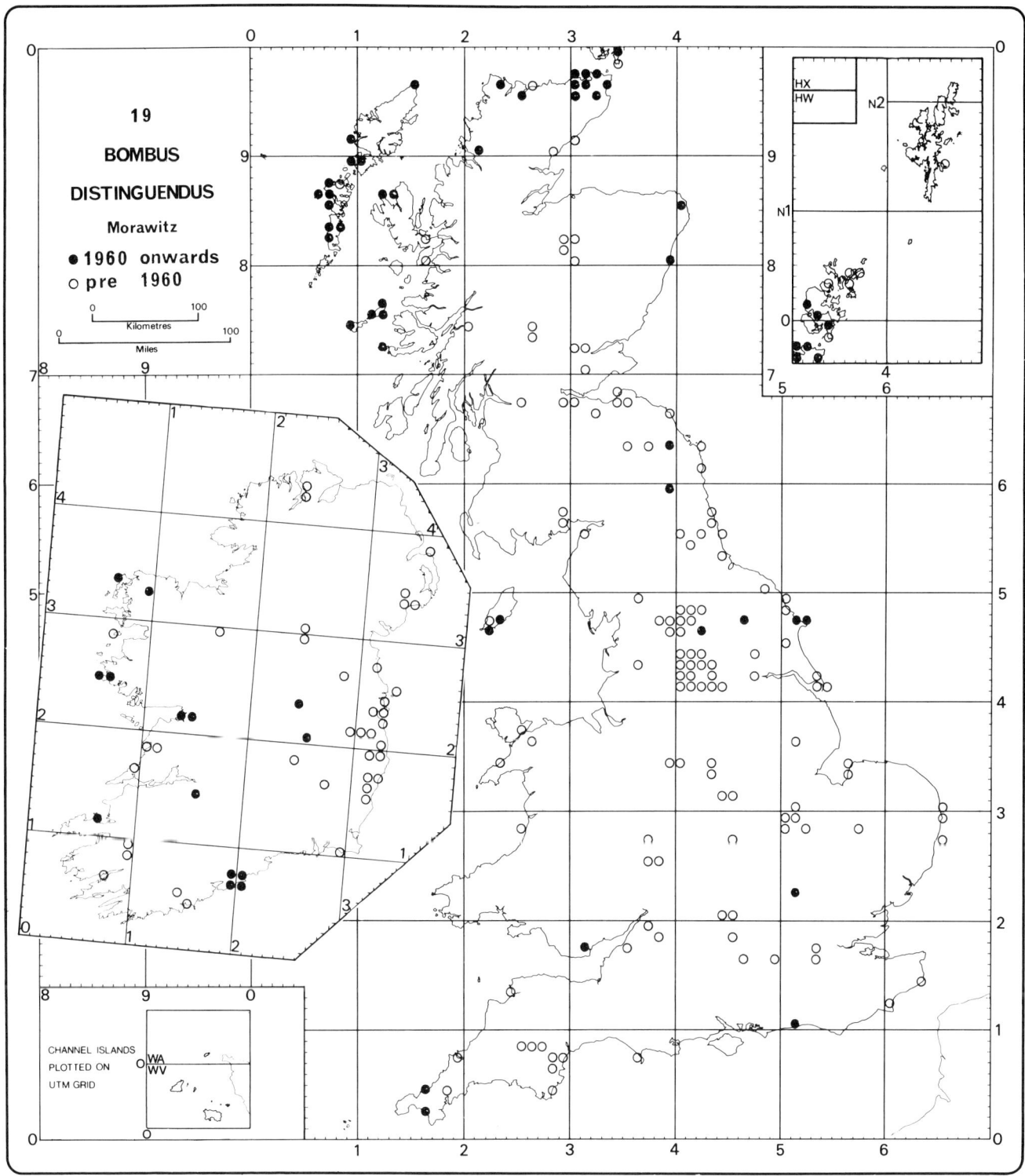

Map 19   Bombus distinguendus *Morawitz occurs mainly in Ireland and north and north-west Scotland; it is very rare elsewhere.*

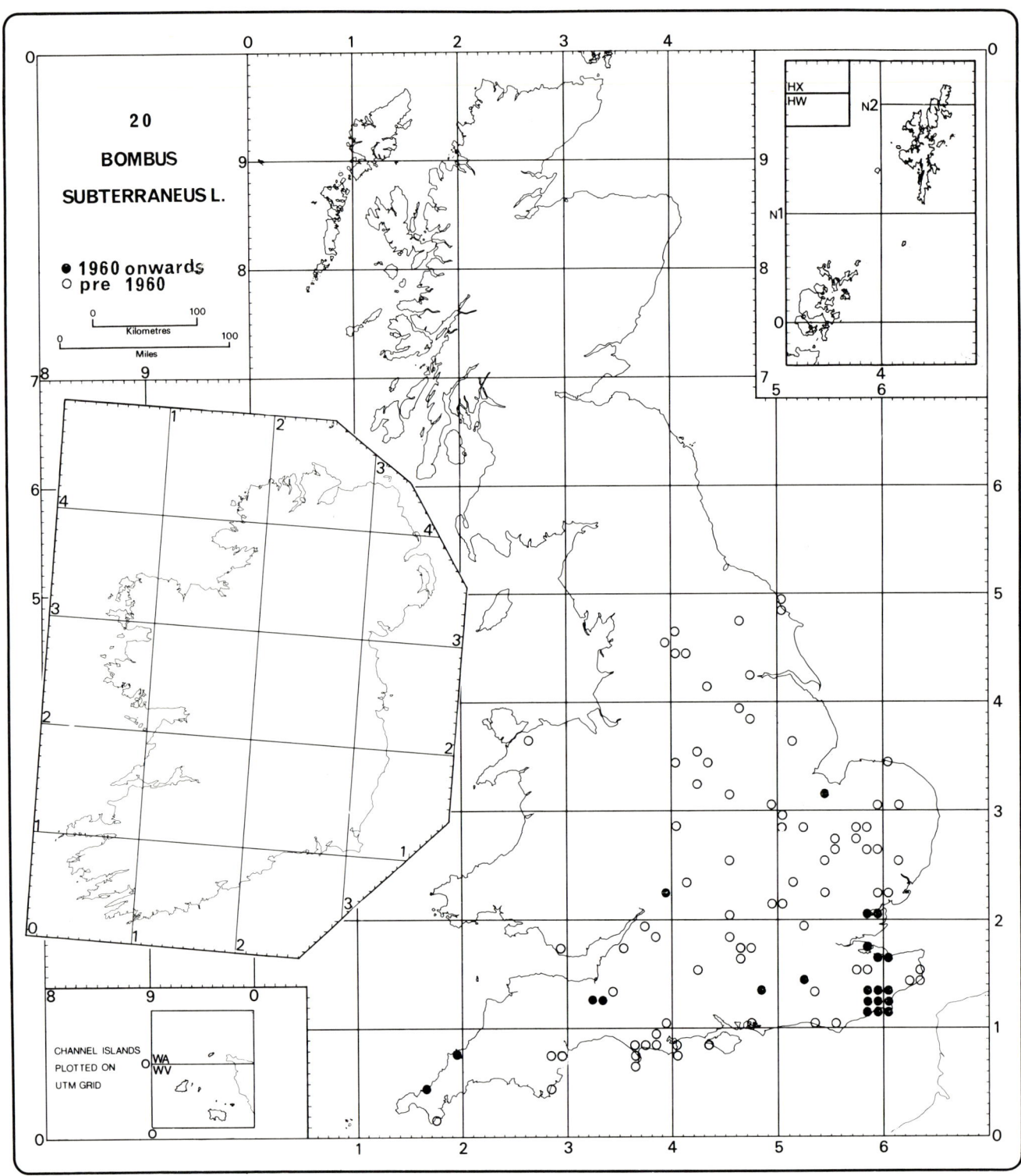

Map 20   Bombus subterraneus *(L.), closely related to* B. distinguendus, *has a southern distribution, and even there it is now rare and very local.*

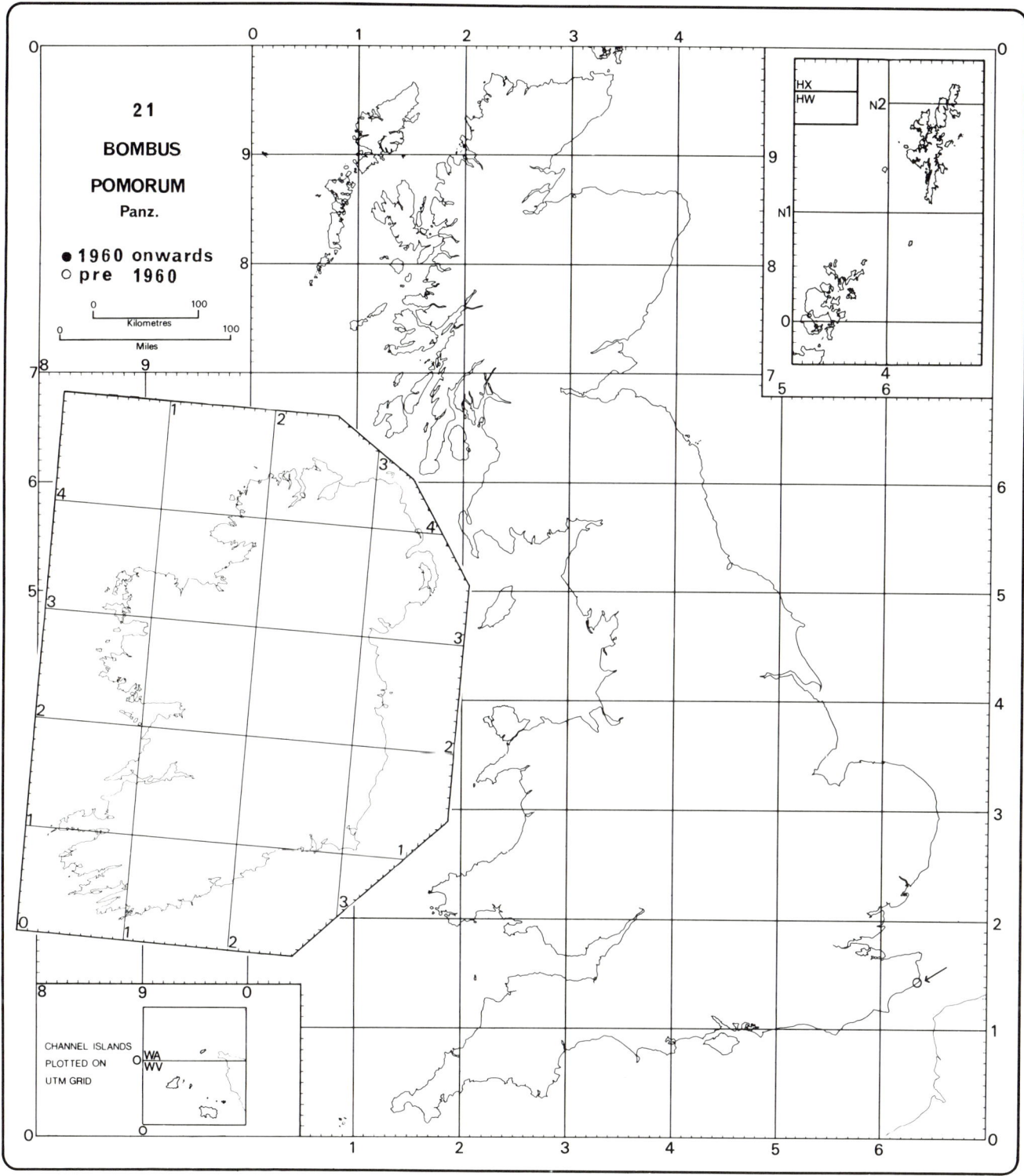

Map 21 Bombus pomorum *(Panzer) has not been recorded since 1864 in Britain (or Ireland) and is presumably now extinct here.*

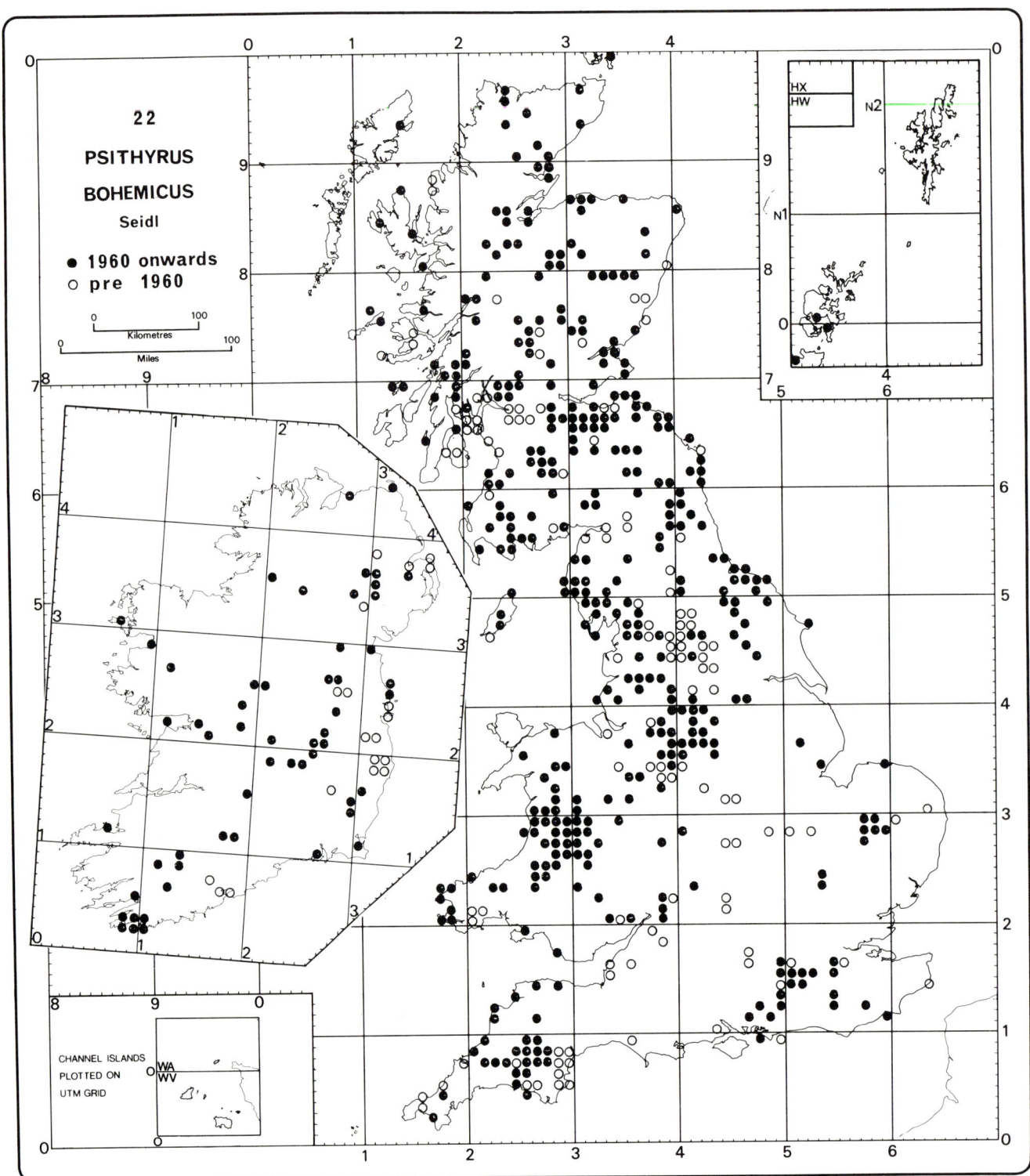

Map 22 Psithyrus bohemicus *(Seidl), an inquiline of* Bombus lucorum, *is a common species particularly in the north and west of Britain.*

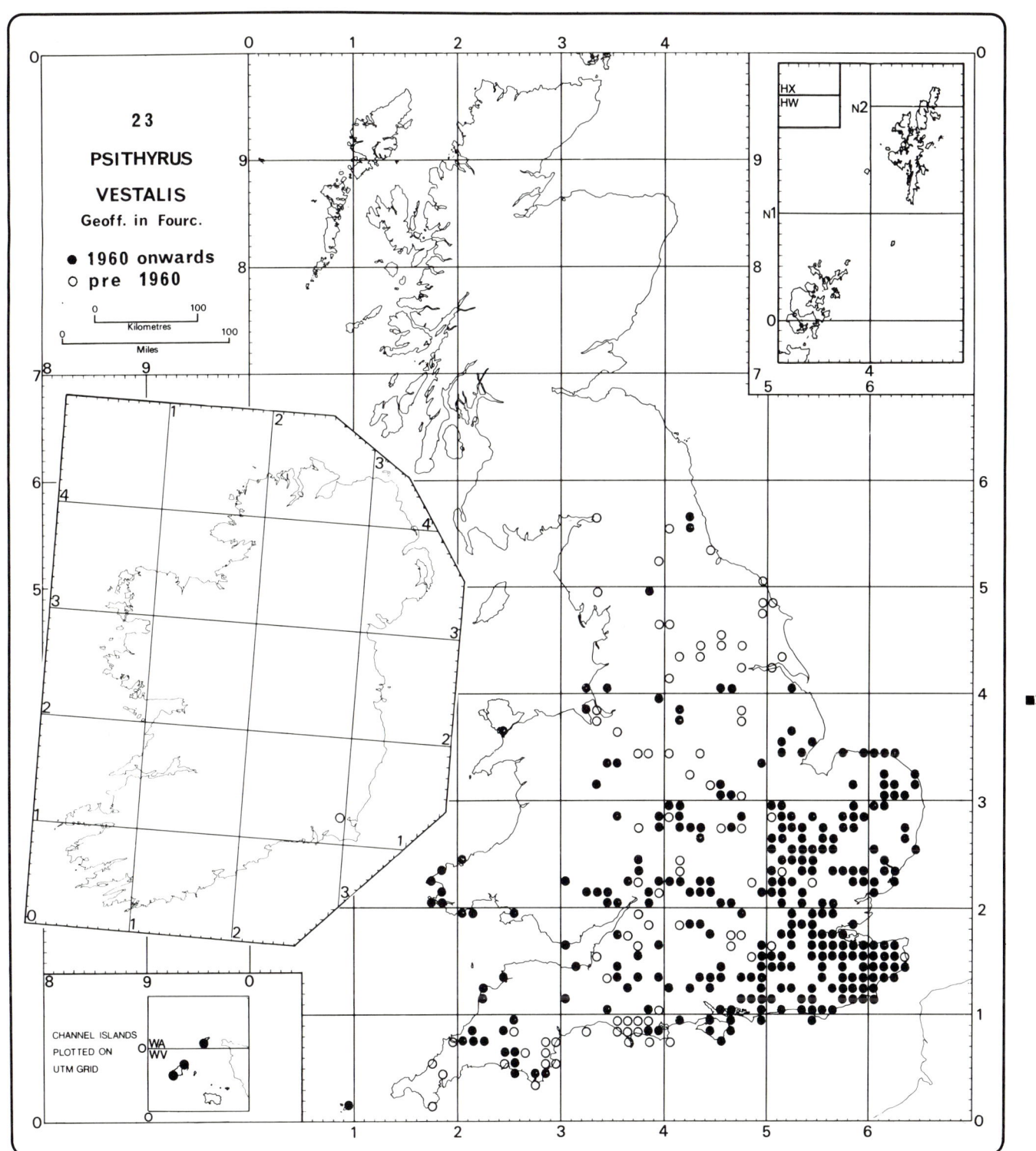

Map 23 Psithyrus vestalis *(Geoff.* in *Fourc.)*, an inquiline of Bombus terrestris, *closely follows the distribution of its host, except in Scotland and Ireland where no specimens were found.*

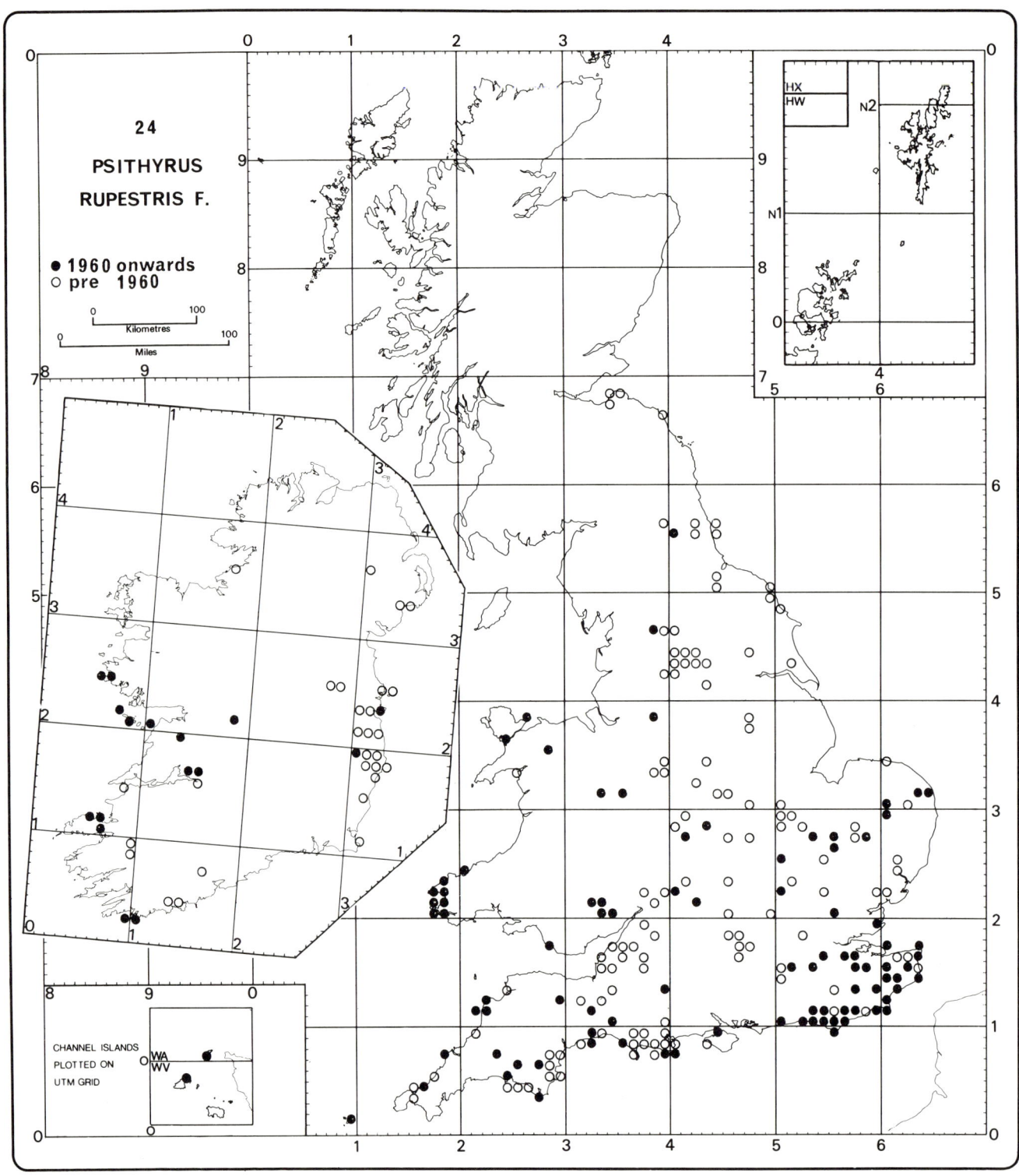

Map 24 Psithyrus rupestris *(Fabr.) is the least often recorded species of* Psithyrus; *it breeds in nests of* Bombus lapidarius.

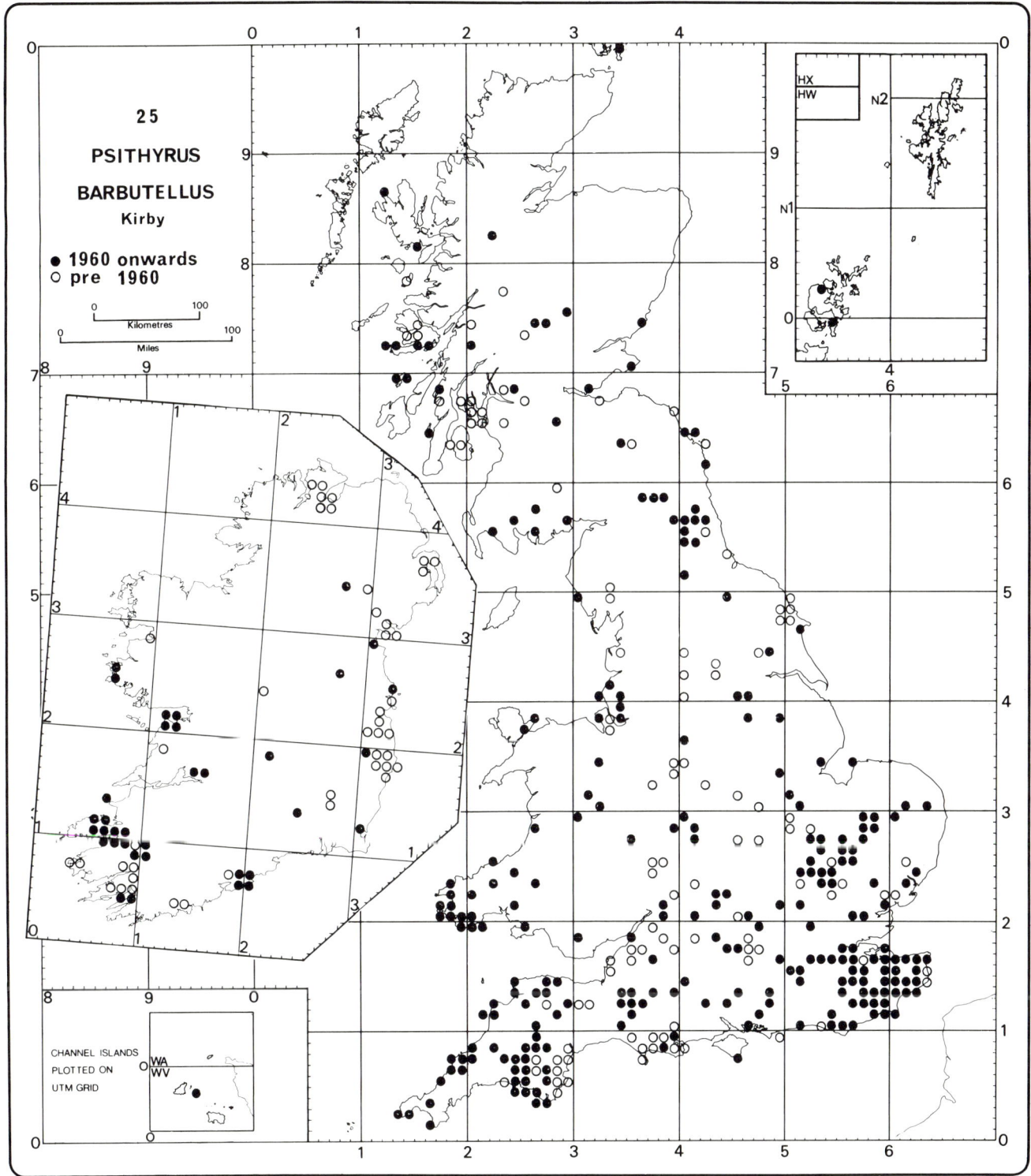

Map 25 Psithyrus barbutellus *(Kirby), an inquiline of* Bombus hortorum *; frequently common.*

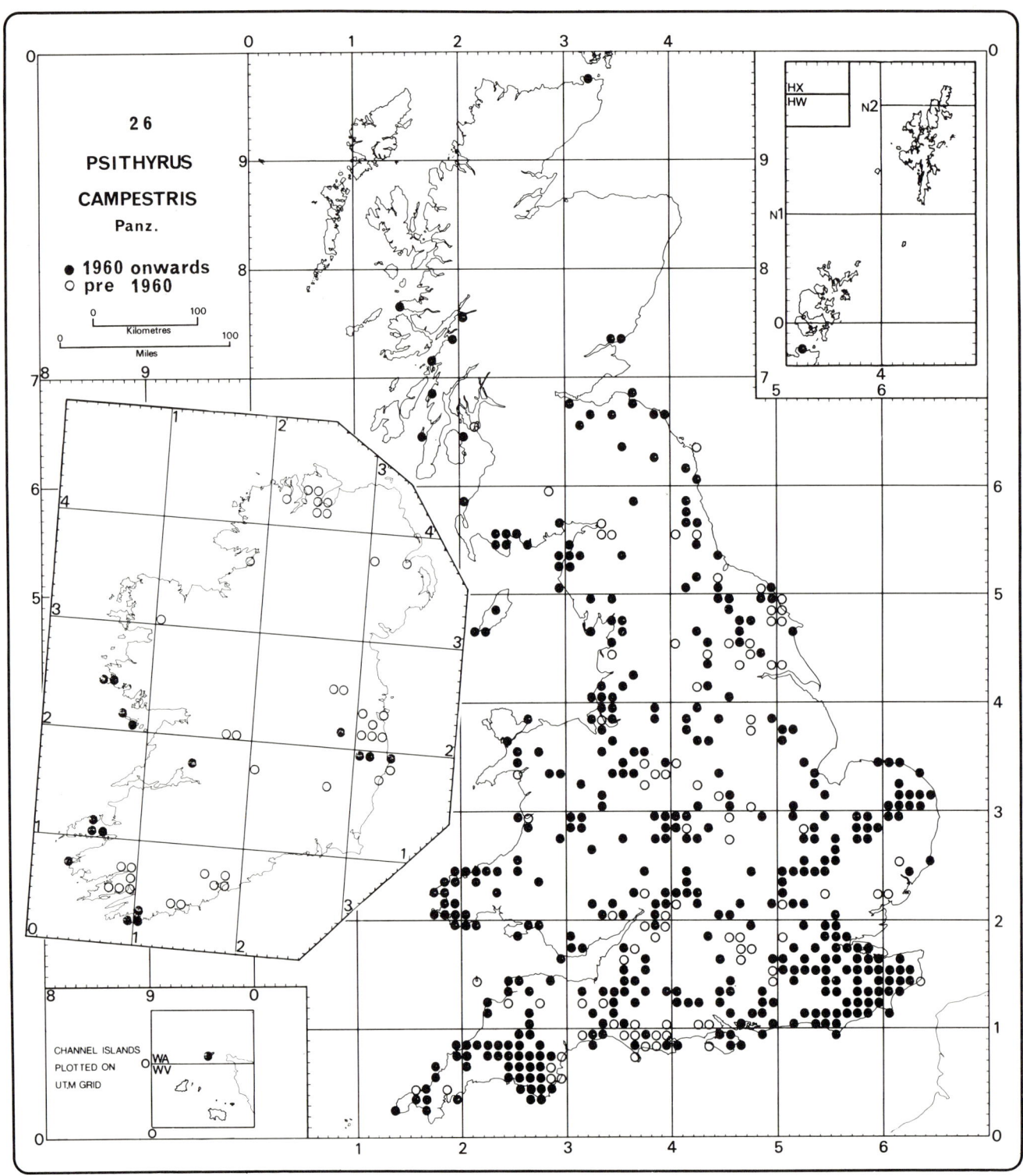

Map 26   Psithyrus campestris *(Panzer), which breeds in nests of* Bombus pascuorum; *frequently common.*

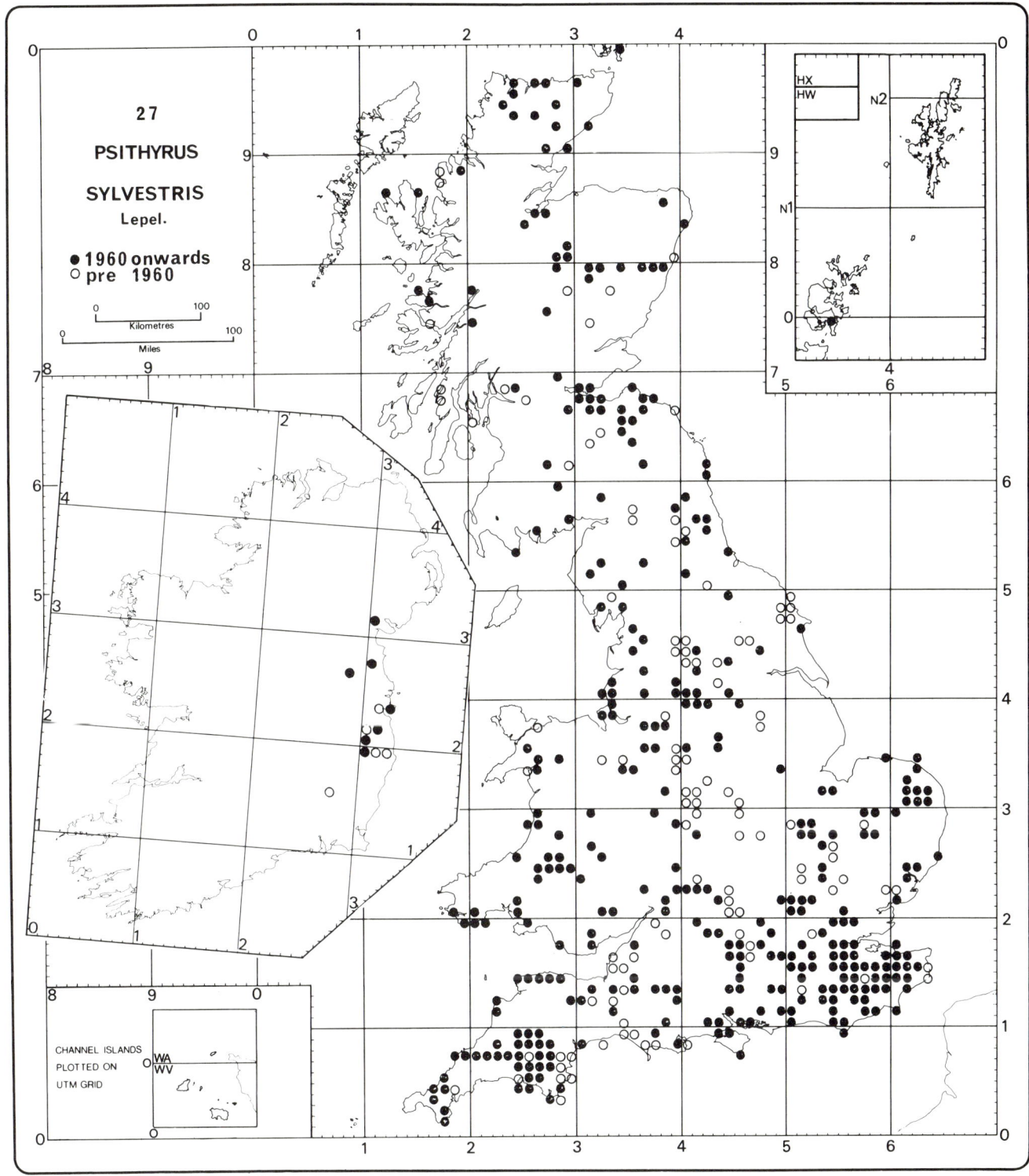

Map 27   Psithyrus sylvestris Lepel., an inquiline of Bombus pratorum and possibly also of B. jonellus; frequently common.

General Bibliography

*Alford, D. V., 1970–72. Bumblebee distribution maps scheme. Guide to the British species. Entomologist's Gaz. *21*: 109–116; *22*: 29–36; 97–102; 229–234; *23*: 17–24; 227–236.

*Alford, D. V., 1971a. Bumble bee distribution maps scheme. Progress report for 1970. *Bee World 52*: 55–56.

*Alford, D. V., ed. (1973). Provisional atlas of the insects of the British Isles. Part 3. Hymenoptera—Apidae (*Bombus: Psithyrus*). Huntingdon; Biological Records Centre.

*Alford, D. V., 1971. Egg laying by bumble bee queens at the beginning of colony development. *Bee World 52*: 11–18, also IBRA Reprint M59.

Alford, D. V., 1975. *Bumblebees.* London; Davis-Poynter.

Alford, D. V., 1978. *The life of the bumblebee.* London; Davis-Poynter.

Crane, E., 1976. The range of human attitudes to bees. *Bee World 57*: 14–18.

Free, J. B. & Butler, C. G., 1959. *Bumblebees.* London; Collins.

*Morgan, P. & Percival, M., 1967. The rearing and management of bumble bees for students of biology. *Bee World 48*: 48–58; 100–109; also IBRA Reprint M48 (which includes a 4 page bibliography on bumblebees).

Plath, O. E., 1934. *Bumblebees and their ways.* New York; Macmillan.

Richards, O. W., 1953. *The social insects.* London; Macdonald.

Sladen, F. W. L., 1912. *The humble-bee, its life history and how to domesticate it.* London; Macmillan.

*Available as IBRA reprints.

Index

The numbers given after each entry are the map numbers.

Bombus

cullumanus	6
distinguendus	19
hortorum	11
humulis	13
jonellus	7
lapidarius	10
lapponicus	8
lucorum	3
magnus	4
muscorum	14
muscorum smithianus	15
pascuorum	16
pomorum	21
pratorum	9
ruderarius	17
ruderatus	12
soroeensis	2
subterraneus	20
sylvarum	18
terrestris	5

Psithyrus

barbutellus	25
bohemicus	22
campestris	26
rupestris	24
sylvestris	27
vestalis	23